Tito Gonzalez

CONTROL OF UNSTABLE SYSTEMS

Tito Gonzalez

CONTROL OF UNSTABLE SYSTEMS

Case study: Quadcopters

ScienciaScripts

Imprint

Any brand names and product names mentioned in this book are subject to trademark, brand or patent protection and are trademarks or registered trademarks of their respective holders. The use of brand names, product names, common names, trade names, product descriptions etc. even without a particular marking in this work is in no way to be construed to mean that such names may be regarded as unrestricted in respect of trademark and brand protection legislation and could thus be used by anyone.

Cover image: www.ingimage.com

This book is a translation from the original published under ISBN 978-3-639-71192-9.

Publisher:
Sciencia Scripts
is a trademark of
Dodo Books Indian Ocean Ltd. and OmniScriptum S.R.L publishing group

120 High Road, East Finchley, London, N2 9ED, United Kingdom
Str. Armeneasca 28/1, office 1, Chisinau MD-2012, Republic of Moldova, Europe
Printed at: see last page
ISBN: 978-620-5-68874-8

Contents

SUMMARY

Aircraft with four or more horizontal rotors have undergone an accelerated evolution in recent years. Systems engineering research groups, control engineering students, and hobbyists in general are working on the implementation of more robust models and controllers, so that more detailed, accurate, and therefore more realistic simulations of quad-rotor behaviour can be worked out. Many authors, thinking ahead to a practical implementation and limited in computational power by the imposed lightness in weight, choose to keep the modelling of the vehicle dynamics relatively simple, which makes it possible to control the quadrotor using classical control techniques of linear systems such as PID, by linearising the dynamics of the device behaviour around an operating point that is usually chosen according to where the device is to hover. However, wide-range flight or more complex manoeuvres require a better performance of the control system, which can only be achieved by the use of non-linear control techniques, as they consider a more general form of vehicle dynamics in all flight zones, but involve a more complete model of the aircraft.

An added value of the developed simulation system is that with the necessary modifications, a simulation platform can be implemented for didactic purposes for teaching the control theory of non-linear systems.

Keywords: Modern Control. Nonlinear Control, Linearization. Optimal Control. Optimisation.

1.1. Problem statement.

The quadcopter is a flying machine which can modify its position and orientation by means of four rotors or propellers which are independently controlled in their rotation and which in sum of their actions provide the performance of this class of aerial vehicles. As such, quadrotors are one of the possible variants in the family of multicopters or devices with several rotating elements for lifting and manoeuvring.

During the 20th century, many attempts were made at practical implementations of manned quadcopters, with the first trials dating back to 1922 in France by Etienne Oemichen [1] and in the United States of America (USA) by George Bothezat [2]. However, during the evolution of the aeronautical multicopter industry, it was necessary to test different rotor adjustment schemes in order to improve manoeuvrability.

In the last decades, thanks to the advances in construction technologies in terms of miniaturisation of electronic components such as: microcontrollers, electric motors, sensors, actuators, and software, the possibility of building small unmanned aerial vehicles (UAVs) but operated by remote control (teleoperated) has greatly increased, which has allowed a growing number of researchers to resume their interest in quadrotor devices due to their ease of construction.

Although currently quadcopters are mainly used as toys or to capture panoramic images (photo, video) thanks to the height they can reach, there is a growing interest at university level to use it as a mechatronic test bench for teaching control theory of non-linear systems, and it is also a very good prospect for application in other areas.

At the university level, the aim is to expand its manoeuvrability in teleoperation, autonomy, and intelligence [3] in order to be used in surveillance or reconnaissance operations, inspection of facilities and structures, terrain mapping [4], search and rescue [5], firefighting, or as an element within a group that must fulfil a general purpose task [6].

On the other hand, quadcopters have advantages such as: high manoeuvrability, relatively low price, and very simple construction. These characteristics give them great potential for use as autonomous robotic devices.

However, there are several operational problems that have to be solved or improved in order to be used in real applications. One of these problems is the fact that as an aerial device and in terms of Aeronautical Engineering it is a highly non-linear and multivariable system, since it has six degrees of freedom (6 GDL), three for the position (x, y, z), three for the orientation (turn, elevation, rotation), and only four actuators, which transforms it into a non-holonomic system. That is, it is an underactuated system [7] or one that has a smaller number of control outputs in relation to the degrees of freedom and therefore cannot move translationally without first rotating about one of its axes.

On the other hand, this operating condition also makes the quadcopter an unstable system by nature (open loop) and very difficult to control due to the non-linear coupling between the actuators and the degrees of freedom [8]. It is because of these conditions that it is a project of great interest in the area of control engineering, since the regulation of the position (3 GDL), the regulation of the orientation (3 GDL), together with the regulation of the linear and angular velocities requires a robust control architecture, since it is essential to coordinate the operation of the four rotors at the same time in order to meet the requirements of the flight manoeuvre.

Despite the difficulties and requirements mentioned above, very good results have been obtained in the control of the quadcopter's operation, as can be seen from the results of the GRASP laboratory of the University of Pennsylvania, USA, and the "Flying Machine Area" project of the University of Zurich, Switzerland.

Therefore, the main motivation for the development of this work is the creation of a simulation and control system with six degrees of freedom by means of the pre-established mathematical model of the quadcopter and with the use of the state variables to create a Linear Quadratic Regulator (LQR) with which it is expected to control the position and orientation of the quadcopter in function of the aerial manoeuvring commands.

In relation to what is established in the previous paragraph, it should be pointed out that, based on the publications reviewed and which will be indicated in more detail in the section corresponding to the background of this research, it is decided to use modern control theory when selecting an optimal controller of the linear quadratic type.

1.2. Formulation of the problem.

For the creation of the simulation system it is necessary to solve several tasks. First of all, several already developed mathematical models have to be evaluated in order to select the one that will be used to represent the quadcopter in the simulations.

In this respect it should be noted that the mathematical modelling of the quadrotor consists of the description of its rigid-body dynamics, the kinematics with respect to the reference frames, both fixed and sailing, and the forces applied to the quadrotor during manoeuvres. In view of the above, it is necessary to point out that there are several methods of determining the model of the craft. Firstly, these can vary in the description of the rigid body dynamics, since this can be determined by means of the Euler equations [9], or by means of the Euler-Newton approximation [10], or by means of the Lagrangian approximation [11]. Secondly, these may vary in the final representation of the kinematics and the direction of the 'z' axis of the artefact reference frame. Thirdly, they may vary in the consideration of how many forces and other disturbance effects must be taken into account.

Once the model to be used has been determined, and based on this, the linear quadratic controller of six degrees of freedom with the use of only four control actions must be designed by means of the state variables of the aircraft, where the interrelation between the mathematical model of the quadcopter and the controller that will carry out the adjustments so that the model follows the manoeuvre orders, will be implemented using the MATLAB mathematical processing software and any of its associated tools.

Once the controller has been coupled with the plant model, a series of manoeuvre actions adjusted to the operating reality of the device will be executed, and from the results of which the considerations for the tuning of the controller will be obtained. Once the adjustments have been made to the linear quadratic controller, the pre-established manoeuvres will be executed again in order to check the correct operation of the unit.

1.3. Objectives.

The following are the expected goals of this research.

1.3.1. General objective.

Develop a simulation system using modern control theory for the operational manipulation of the quadcopter.

1.3.2. Specific objectives.

Determine the mathematical model that will represent the quadcopter.

, Design the linear quadratic controller with six degrees of freedom and four actuators.

, Implement the quadcopter model and the linear quadratic controller.

in the MATLAB program and its associated tools.

, To develop the manoeuvring actions for the tuning of the linear quadratic controller.

, Evaluate the performance of the linear quadratic controller.

1.4. Justification of the Investigation.

The development of this work will allow the creation of a computational simulation system that will facilitate the master students in the areas of applied mathematics (mathematical modelling) and control theory (Electronic Engineering) the strengthening of the acquired knowledge, since through the simulation they will experiment with a system that represents a risky device without the risk of damage to the air vehicle during the manoeuvres and tests in the control techniques.

An added value of the developed simulation system is that with appropriate modifications it can be implemented as a didactic test platform for teaching the control theory of non-linear systems.

If at a later date it becomes possible to obtain the necessary financial resources, all the equipment and components necessary for the development and implementation of a laboratory where the validity of the theoretical concepts studied in this work can be verified by experimentation.

Finally, and in terms of the future, a possible application that could be obtained with the development of this master's degree work is to enhance the security and safety systems of the UNET with a set of drones with which a better surveillance system of the university campus and its surroundings can be developed.

2.1.Background to the investigation.

Aircraft with four or more horizontal rotors have been the subject of accelerated evolution in recent years. Research groups, engineering students, and hobbyists in general are working on the implementation of more robust models and controllers, so that more detailed, accurate, and therefore more realistic simulations of the behaviour of quadrotors can be worked out.

From the various reviews it can be seen that many authors, thinking ahead to a practical implementation and limited in computing power by the imposed lightness in weight, opt to keep the modelling of the vehicle dynamics relatively simple, which makes it possible to control the quadrotor using classical control techniques of linear systems such as PID (Proportional + Integral + Derivative) as indicated in the works of Hoffmann et al [12], Goela et al [13], Erginer et al [14], Jategaonkar [15], Cowling et al [16], Bouabdallah et al [17], and Pounds et al [18], by linearising the behavioural dynamics of the device around an operating point that is usually chosen according to where the device is to hover. However, wide range flight or more complex manoeuvres require a better performance of the control system, which can only be achieved by the use of non-linear control techniques, as they consider a more general form of vehicle dynamics in all flight zones, but involve a more complete model of the aircraft.

In terms of non-linear control methods, work has been carried out with techniques such as: Linear Quadratic Regulator (LQR) optima as reported in the works of Cowling et al [19], Minh et al [20], Al-Younes et al [21], and Bouabdallah [22], in addition to various adaptive and semi-adaptive schemes as reported in the work of Mellinger et al [23], or using the technique of Recursive Control (Backstepping) or recursion as shown in the works of Altug et al [24], Madani et al [25], Madani et al [26], and Bouabdallah et al [27], or Sliding Mode control as shown in the work of Xu et al [28], and Lee et al [29], or feedback linearisation as reported in the work of Altug et al [24], Mokhtari et al [30], and Das et al [31], through to work with dynamic model inversion as reported by Hehn et al [32], and Das et al [31]. Works from which it can be inferred that these are effective techniques for the control of a frame that must perform piloting manoeuvres. In particular, feedback linearisation seems to show very favourable expectations for the control of quadcopter-type air vehicles, since there is a linearisation structure that can represent the quadcopter dynamics in terms of the orientation and height within the inner loop, and the position within the outer loop.

In the work presented, various forms of feedback from the various sensors carried by the aircraft are assumed, such as inertial measurement units (IMU), linear and angular acceleration measurement units, infrared and sonar devices to measure the distance to obstacles in the flight path, normal and infrared cameras, and global positioning system (GPS) units to determine spatial location. Typically, when the proposal is implemented experimentally, it is done using the aircraft's electronic circuitry, and in some cases, infrastructure external to the aircraft but linked wirelessly to a ground station is used for control planning or trajectory generation. Unfortunately, many of the proposals and control schemes are not implemented experimentally and only the mathematical justification or simulations with analysis of the results are provided, and when experimental implementation is reached, a comparison between simulated and obtained under identical conditions is usually not established for the purpose of model verification and validation.

Returning to the idea of mathematical manipulation, all of the control techniques suggested above require a complete knowledge of the model and system parameters, but numerical errors in the parameter values used can lead to significant deterioration in controller performance. In addition, unmodelled variations in system parameters such as mass and hence inertia during flight manoeuvres can cause significant errors in stabilisation to occur in the case of linearised systems and classical control techniques.

On the other hand, the need for an accurate nonlinear model of the quadrotor dynamics can be satisfied by using adaptive methods that can react to perturbation and correct errors in the model parameter estimates by adjusting the parameters under adverse conditions. In that sense, methods such as Model Reference Adaptive Control (MRAC) proposed by Whitehead et al [33] could be used for that purpose.

However, with respect to many of the linearised control methods, the achievable trajectory of the vehicle is constrained by the linearisation utilisation itself. In the work developed by Huang et al [34] an adaptive recursive control method is suggested which can be extended to include the inertia parameters in the adaptation law developed by Zeng et al [35].

Recent work in the search for the operational autonomy of a quadrotor device used indirect adaptive least squares based methods for the vehicle mass problem such as that proposed by Mellinger et al [23]. However, all indirect methods based on differences between expected and actual plant outputs correct for errors in the parameters, but do not do so explicitly in the model parameters, as do the direct adaptive methods suggested by Craig et al [36] for robotic mechanical manipulators.

On another note, it is important to keep in mind that progress in sensor technology, information processors, and actuator integration has made possible the existence of really small flying robots. Bouabdallah et al [37] describe a possible approach in the sense of developing a micro quadrotor for indoor use which they call "OS4".

In another paper, Bouabdallah et al [17] present the results obtained using the classical PID (Proportional + Integral + Derivative) controller and a Linear Quadratic (LQ) controller when implemented in the OS4, controllers based on a more developed model than the one originally used [37]. The results obtained with the quadratic controller reveal that it tends to be problematic as it is difficult to find the values of the matrices that satisfy the stability condition of the robotic assembly. However, the classical PID controller approach performs favourably compared to the LQ controller due to the simplicity of the method chosen to deal with the model uncertainties, as long as the airburst disturbances are kept at zero or at a minimum.

In the work presented by Becker et al [38], a computational tool specifically designed to simulate the dynamics of the OS4 is developed and implemented in MATLAB via Simulink, which allows the testing of the response of various sensors and filters for the generated signals, as well as various control techniques. This simulation environment was specifically used, in this case, to test the obstacle avoidance capability using ultrasonic sensors. Obviously, before proceeding to the experimental implementation and testing on the OS4.

With the aim of designing a practical quadrotor helicopter, Pounds et al [39] created the "X-4 Flyer Mark II" which is a structurally robust quadrotor platform with a custom-built airframe and avionics and a payload capacity of up to 4 kg. It has the particularity that the design and construction of the propellers is aero-elastic and the performance results of this feature are reported. As such, the quadrotor includes spring-loaded balancing in the rotors, which allows adjustment of the blade's beating characteristics. This effect is also included in the dynamic modelling of the device developed in MATLAB and from whose simulations it can be seen that the configuration and inverse operation of the propellers is beneficial for the flight of the quadrotor.

This same team of researchers presents in later work [40] the analysis of the dynamics of the vehicle posture, which allows the tuning of the mechanical design to improve the response in the rejection of disturbances and the sensitivity to control. For the purpose of velocity stance control, a linear SISO (Single In Single Out) controller was implemented to stabilise the dominant and decoupled modes of lift (Pitcl) and roll (Roll) when using a perturbed input model to estimate the plant performance (quadrotor). The plant model also implements the flapping of the propeller blades, results which are in agreement with the experimental values obtained from the "X4 Flyer" and which indicate that the blade flapping characteristic introduces very significant flight stabilising effects on the vehicle. Overall, the results show that the propeller flap compensation allows for low speed yaw control of the rotors.

On the other hand, Mokltari et al [41] present a mixed linearisation by robust feedback with GH4 linear controller applied to the quadrotor nonlinearities. Actuator saturation and constraints on the state space outputs were implemented to analyse the worst-case scenario in the design of the control laws. In addition, the controller behaviour was checked for subjugated uncertainties in the parameters, external disturbances, and noise in the sensor measurements. The simulations show that the overall system behaviour becomes robust when the weighting values of the functions are carefully selected.

Artificial vision has also proven to be a useful tool for the spatial control of quadrotors. Tournier et al [42] present the estimation and control of a quadrotor based on vision by means of a simple camera (Webcam) in relation to an aerial vehicle using Moire patterns. The purpose of this research was the acquisition of six estimated degrees of freedom, which are essential for the operation of approaching aerial vehicles on approach to each other or on approach to the landing pad. Test results indicate that it is possible to use this system to autonomously maintain the hovering approach position of the aircraft as long as the target location remains within the field of view of the camera.

Taking into account that current quadrotor designs generally only consider nominal operating conditions in the design of control strategies, Hoffman et al [12] address the problem that arises when the aircraft's operation deviates significantly from its hover point due to wind gust disturbances. In particular, they investigated three different effects on the aerodynamics of the quadrotor and their relationship to the vehicle's speed, angle of attack, and airframe design. The first effect studied was the way in which the total thrust varies and not only with variations in the power of the engines that power the rotors. This included the effect of the rotor exit jet in relation to the angle of attack of the blade with respect to the free blast. The second phenomenon studied was the "propeller blade flapping", which results from the different velocities of the air entering or being directed towards the propeller due to its rotation and which is produced by the advance and retreat of the blade. This phenomenon induces roll and pitch moments in the rotor hub as well as deflections in the thrust vector. Finally, the interference effect caused by the components that make up the body of the vehicle and that are close to the sliding flow from the propellers was investigated as it is mainly responsible for the instabilities in the net thrust and difficulties in the tracking of the navigation posture.

The full extent of all these aerodynamic effects were observed in practice by means of the STARMAC II quadrotor, which has autonomous control over its attitude and lift, demonstrating in the results obtained that mathematical models of the aircraft and derived control techniques are inadequate for accurate trajectory tracking at significant or high speeds and in uncontrolled environments.

On the other hand, Courses et al [43], provide a fully dynamic mathematical model of a commercially available quadrotor and present its performance at low altitude under a sliding mode control scheme, which is a technique known for its robustness in maintaining the operating point despite disturbances and variations during operation. The plant was modelled nonlinearly and with strongly coupled state variables. Assumptions made included: unsaturated domains, minimal climatic disturbances, actuating elements with a reasonably fast response speed. Simulations showed that the control algorithm successfully piloted the system according to the programmed trajectory with tracking in the control signals.

Katie Miller [44] seeks to design control laws for a quadrotor helicopter that traverses a predetermined route by defining points of interest, more specifically, using a PD controller for position control, and a PID controller for rotation control that defines the new direction once the point of interest is reached. The control laws were developed from the dynamic model of the quadrotor, which was obtained by linearising its non-linear equations of motion. The behaviour of these control laws was studied under the presence of model uncertainties and external disturbances such as wind gusts. The results show that while the linearisation control laws are adequate under conditions of almost no perturbation, in the presence of environmental disturbances the performance significantly deteriorates.

Gjioni et al [45] designed a quadrotor drone for the sixth annual unmanned aerial systems competition, where this quadrotor uses a GPS (Global Positioning System) for route tracking and altitude control, also uses a camera and has computational capabilities thanks to a dual processor, making it capable of autonomous navigation and information exchange with the ground station. Posture control is provided by four PID controllers.

(lift, pitch, yaw, yaw, and acceleration control) where for each case the tuning of the controllers is performed experimentally. In relation to the practical results, it should be noted that during the competition the aircraft encountered severe crosswind turbulence, but the controllers performed exceptionally well as they were able to stabilise the system during the flight.

2.2. Theoretical Basis.

The following are the minimum concepts necessary for the understanding of the mathematical formalities of the research project.

2.2.1. A bit of history.

The story as such begins in the early 20th century, when Charles Richet, a French academic and scientist, built a small prototype of an unmanned helicopter [46]. Although this attempt was not a success, Louis Breguet, a student of Richet's was inspired by his professor's tests, so that by 1906 Louis and his brother, Jacques Breguet, began construction of their first quadrotor. Louis carried out many tests on the blade profile for the propeller design, so that he gained a basic understanding of the requirements necessary to achieve vertical flight.

In 1907 they managed to finish the construction of a device for vertical flight that they called: Breguet-Richet Gyroplane No 1, which was a quadrotor with double propellers of 8.1 metres in diameter each, a weight of 578 kg (including the two pilots), and a single internal combustion engine of 50 HP (37.3 kW) that powered the propellers by means of a belt transmission. One of the curiosities of these early trials is that none of the pilots had any idea how to control the stance of the craft, only the main aspect was to be able to perform a vertical take-off, hover over a particular position, and then make a vertical landing.

Another pioneer of aeronautics, the engineer Etienne Oemichen also experimented by 1920 with designs using rotary wings. In all, he designed six different vertical lift machines. The first model failed in its attempt to rise from the ground but, because of Oemichen's determination, he decided to add hydrogen-filled balloons to provide the craft with both stability and upward thrust. The second craft, the Oemichen No2, had four rotors and eight propellers supported by a cross-shaped steel tube frame. Five of the propellers were for lifting and stabilising the craft, one propeller for steering, and two propellers for forward propulsion.

Although very rudimentary, this machine achieved a considerable degree of stability and controllability so that by the mid-1920s it made over a thousand test flights without major problems. It was the first aircraft to stay in the air for several minutes without any ground support, and by May 14th this machine was airborne for 14 minutes and made a flight of over 1600 metres.

Not surprisingly, the US Army took an interest in vertical ascent machines, so that by 1921 they contracted Dr. George de Bothezat and Ivan Jerome to develop a device for use by the US Army Air Corps. The result was a 1678 kg structure with 9-metre arms and four rotors with six-bladed propellers 8.1 metres in diameter. One of the conditions of the army contract required that the constructed craft be able to hover at a height of 100 metres, but the highest height achieved by the device was only 5 metres. At the end of the project

Bothezat demonstrated that the vehicle could be very stable, however, multiple technical problems with the technology available at the time put paid to the continuation of the project.

Much later, in 1956, a prototype quadrotor helicopter was built and called the "Convertawings Model A". It was designed for both civil and military applications. As such, it was controlled by varying the thrust of the different rotors and although it was a success for the time in terms of vertical take-off and landing, it had the disadvantage that all manoeuvres had to be performed in forward flight, which was not a significant limitation due to the possibility of stationary manoeuvring. The project was abandoned mainly because of the lack of demand for this vehicle.

Figure 1 Convertawings Model A quadcopter.

From that date to the present day much development has been achieved with single propeller and tail rotor helicopters, however, recently there has been an increased interest in quadrotor design thanks, on the one hand, to the availability of very low cost drones, and on the other hand, the interest of large aerospace companies in the construction of aircraft with two or more propellers due to the availability of greater thrust for the transport of bulky and heavy personnel or equipment.

In this regard, Bell Aircraft Corporation is working on a tiltrotor quadrotor with a tiltrotor propeller that surpasses the Bell-Boeing V-22 Osprey in versatility. See Figure 2. It would be capable of carrying large payloads in the normal way, achieve high speed, and be able to perform both take-offs and landings on a runway even if it cannot do so vertically (VTOL, Vertical Take-Off and Landing). Obviously, many of these systems are directly derived from the V-22 with the exception of the number of turbo propeller engines, and the design of the wing that supports the rotor tip has a number of improvements in terms of cruise speed flight, such that high-speed fuel economy is achieved [47] [48] [49]. See Figure 3.

Another recent and famous quadrotor design is the Moller Skycar. See Figure 4. It is a prototype for a personal "Flying Car" with VTOL aircraft characteristics.

This aerial vehicle has four tilting turbo fans which allows for efficient and safe operation at very low speeds. However, despite the publicity, it has been heavily criticised as the only flight demonstrations have been vertical take-off and hovering with the Skycar tethered to a crane arm at a higher altitude [50].

Figure 2 V-22 Osprey.

Figure 3 Conceptual design of the Bell Aircraft Corporation's "QTR, V-44".

Figure 4 The Moller Skycar during a take-off test.

The inventor of the Skycar, Paul Moller, has tried to sell this airship at auction without success. He has now focused his work on the Skycar's precursor, the

"Volantor M200G", a flying saucer-like hovercraft whose latest variant uses eight computer-controlled turbo fans and is capable of hovering up to three metres above the ground, see Figure 5. This maximum height limitation is imposed on the computer control systems due to US Federal Aviation Administration (FAA) regulations which state that any craft flying more than 3 metres above the ground falls under the laws regulating aeroplanes [51].

Figure 5 Moller Volantor M200X.

2.2.2. Principles of quadrotor operation.

In a quadrotor, each rotor and its associated propeller is responsible for a certain amount of thrust on the vertical axis (u_z , Figure 6) and torque about its centre of rotation in the horizontal plane of location (plane u_x , u_y , Figure 6), as well as a certain drag force opposite to the direction of flight due to the effects of the speed of travel on the airfoil.

As for the four propellers associated with each rotor, they are not all the same. In fact, they are divided into two pairs, two pushing for clockwise rotation and two pulling for counter-clockwise rotation, i.e. one pair rotates and the other pair counter-rotates. As a consequence, the resulting torque in the horizontal plane of location can be zero if all the propellers rotate with identical angular velocity, which allows this type of aerial vehicle to remain in the same position and above its centre of gravity. See Figure 6.

In order to determine the orientation of the aeroplane around its centre of mass, aerospace engineers usually define three dynamic parameters identified as the angles: Yaw, Pitch, and Roll, which are shown in Figure 6 following the right hand rule for the definition of the axes. These parameters are very useful because the forces used to control the aeroplane act on the centre of mass so that as the vehicle is lifted it describes turns, pitch, and roll.

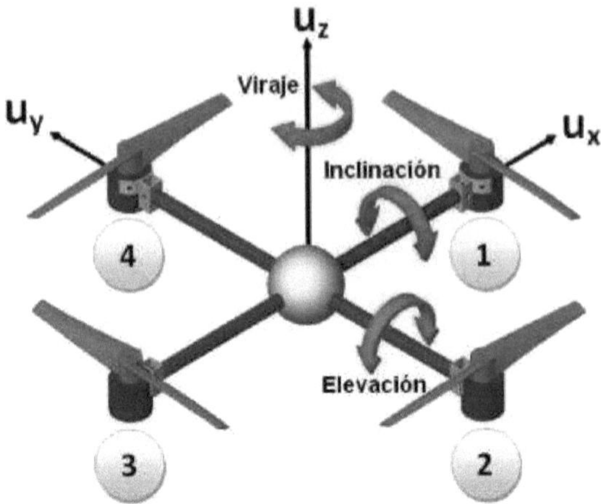

Figure 6 Rotating angles of the frame.

As for the quadrotor, as shown in Figure 7, changes in the value of the elevation angle are achieved by varying the rotational speed between propellers 1 and 3, keeping the speed of propellers 2 and 4 constant, resulting in backward translation, Figure 7a, or forward translation, Figure 7b.

If the same is done with the rotational speed of propellers 2 and 4, keeping the speed of propellers 1 and 3 constant, the change in the angle of inclination is achieved which results in a translational movement to the left, Figure 7c, and to the right, Figure 7d.

If the rotational speed of all four propellers is increased equally, Figure 7e, the artefact rises because of the increase in thrust, and if the rotational speed of all four propellers is decreased equally, Figure 7f, the artefact falls because of the reduction in net thrust.

On the other hand, changes in yaw angle are achieved by means of an imbalance in the balance of aerodynamic torques, i.e. the rotational speed of propellers 1 and 3 is increased in the same proportion while the speed of propellers 2 and 4 is reduced in the same proportion, Figure 7g, while maintaining a constant upward thrust to achieve a left turn.

In the same idea, if the rotational speed of propellers 2 and 4 is increased in the same proportion while the speed of propellers 1 and 3 is reduced in the same proportion, Figure 7h, the device floats horizontally but now turns to the right.

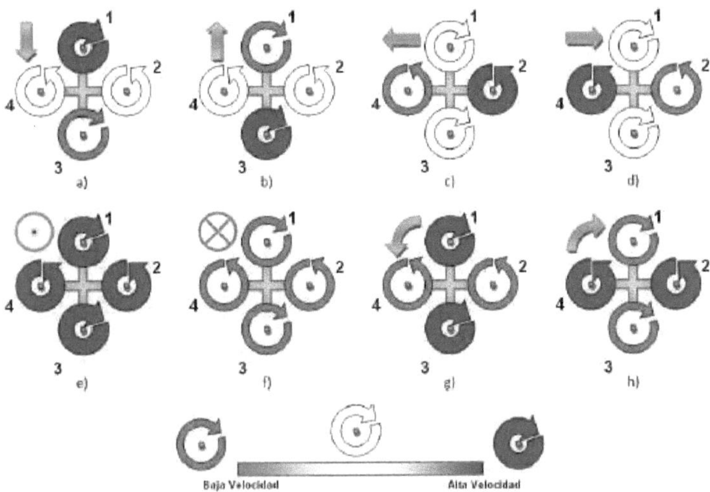

Figure 7 Some actions for simple Quadrotor movement.

Therefore, changing any of these three angles, either individually or together, makes it possible for the quadrotor to manoeuvre in any direction.

As can be inferred from Figure 7 as a whole, quadrotor helicopters do not require tail rotor nor do they require control of the angle of attack of the rotor blades for aerial manoeuvres, as this functionality of independently controlling the rotation speed of each of the propellers, apart from increasing or decreasing the vertical thrust, allows changing the torque applied to the main body, thus eliminating the need for additional torque to control turns.

This feature of using four, or more, independent rotors allows for some important advantages such as: decreasing the mechanical complexity; increasing the robustness of the main frame; simplifying somewhat the modelling of the rotor, if it is assumed that the thrust force exerted by each rotor and its torque vector has an effectively constant orientation with respect to the fuselage of the vehicle and ignoring, of course, the aerodynamic effect of the rotor blade flapping.

In addition, the existence of two or more rotors makes it possible to use the available power for climbing or lifting, which allows the weight of the payload to be significant in relation to the weight of the vehicle. The latter makes this class of vehicles ideal platforms for applications involving video cameras, sensing instruments, transport of materials and tools, autonomous surveillance applications, exploration of relatively open spaces or hazardous areas.

2.2.3. Coordinate system.

When a quadcopter is navigating in three-dimensional space, it does so in two different coordinate systems. One of these systems is the velocity coordinate system, with supra-index "B", which is affected by the thrust developed by the rotors. The other is the navigation frame with sub-index 'E', where forces such as gravitational forces are represented in its influence.

In this sense, in order to develop the mathematical model of the quadcopter, it is essential to first define the coordinate systems with which the movements of the aerial vehicle can be described. These two systems according to Beard [52] are:

1. Inertial or navigational coordinate system (fixed on the earth, F).^{E}

2. Mobile or body co-ordinate system (fixed on the vehicle, $_{FB}$).

Where some physical properties of the quadcopter such as pitch, elevation, yaw, angular velocity can be measured in F_E, while other properties such as linear acceleration can be quantified in F^B .

F^E is the inertial navigation coordinate system and is the reference point for the device. This reference coordinate system can be placed wherever it is convenient, but once set it must be fixed as soon as the quadrotor begins to move. One of the considerations for the reference coordinate system is that the curvature of the earth is ignored. That is, the reference system is placed on a plane, and the quadcopter has a limited projection area on this plane to navigate in space, where this assumption does not affect any possible outcome.

As such, F^E is governed by the right hand rule in both the definition of the axes and the vector product, and where the positive direction of the $Z\text{-}axis_E$ is in the direction from the ground, opposite to the effect of the acceleration of gravity (unit vector u_z , Figure 6). Using this same system, the position of the quadcopter $^{\zeta}$ and its posture are defined. $^{\eta.}$

On the other hand, F^B is the coordinate system that is linked and aligned with the arms that support the vehicle's rotors in such a way that its origin is the centre of mass of the vehicle, where the positive direction of the X axis (X_B , unit vector u_x , Figure 6) goes through thruster 1 which is located towards the front side of the quadcopter. With the same consideration, the positive direction of the Y axis (Y_B, unit vector u_y , Figure 6) goes through thruster 4 which is located towards the left side of the quadcopter.

The $Z\text{-}axis_B$ is perpendicular to the $X\text{-}axis_B$ and $Y\text{-}axis_B$ and its positive direction is in the direction of the thrust forces of the thrusters. In this coordinate system the linear velocities $_{VB}$, angular velocities ω^B , forces f $_B$ and torques are determined. τ^B.

In the same idea, the quadcopter position is defined according to the vector $<\!f,$

as indicated in (1) between the origins of F^E and F^B. See Figure 8, where:

$$\xi = \left[X_{E'},\ Y_{E'},\ Z_{E'} \right]^{T} \tag{1}$$

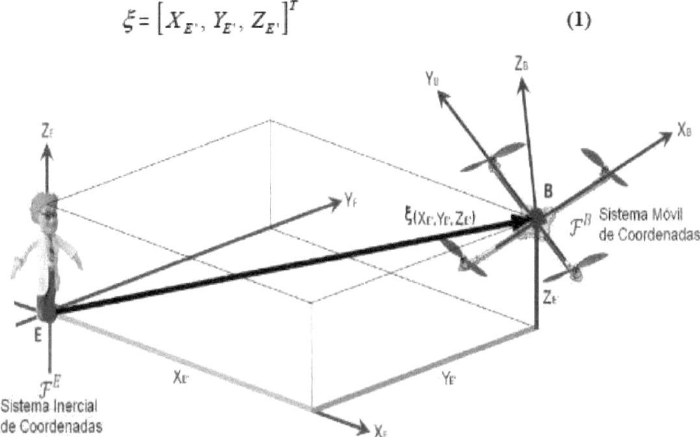

Figure 8 Reference coordinate systems.

while the quadcopter's attitude, n, is defined according to the orientation in FB with respect to the system F^E as indicated in (2). In order to obtain the orientation of the quadcopter in terms of its angles, the artefact coordinate system has to be linked to the navigation coordinate system and the way to achieve this is with the implementation of Euler angles. That is, the orientation is defined according to three consecutive rotations around the coordinate axis system $F^{E'}$, where the values of inclination $^{(\varphi)}$, elevation $^{(\theta)}$, yaw $^{(\psi)}$ are defined. See Figure 9, where:

$$\eta = \left[\phi,\ \theta,\ \psi \right]^{T} \tag{2}$$

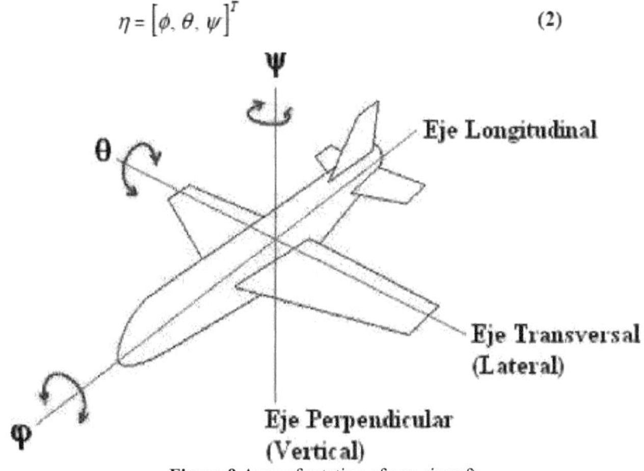

Figure 9 Axes of rotation of any aircraft.

As for the equations of motion of the vehicle, it is more appropriate to formulate them in the system defined by F^B for several reasons, among them: the inertia matrix of the system is time invariant, the system equations are simplified because of the symmetry in the quadcopter frame, the sensoric measurements are easily convertible to the reference coordinate system F^B , thus simplifying the variables in the control equations.

2.2.4. Euler angles.

Euler angles are three angles introduced by Leonard Euler to describe the orientation of a rigid body in space with respect to a reference point in the same space. In general, to describe the orientation of an object in Euclidean space, three parameters Z Y X are required, which can be given in various ways, although in avionics the Euler angles are preferred [53]. See Figure 9.

They can also be used to describe the orientation of one frame of reference, or coordinate system, relative to another and its transformation from the coordinates of a point in one frame A to the coordinates of the same spatial point in another frame of reference, B. Euler angles are typically denoted as:

$$\phi \in \left]-\pi,\pi\right], \; \theta \in \left]\frac{-\pi}{2},\frac{\pi}{2}\right[, \; \psi \in \left]-\pi,\pi\right].$$

A very important aspect of the Euler angles is that they represent a sequence of three elementary rotations on different axes, so that any orientation can be obtained by the composition of these three elementary rotations, as indicated in (3), (4), and (5). Obviously, these rotations start at a commonly known orientation, and each of the rotation matrices [54] is indicated below.

$$R_x(\phi) = \begin{bmatrix} 1 & 0 & 0 \\ 0 & \cos(\phi) & -\operatorname{sen}(\phi) \\ 0 & \operatorname{sen}(\phi) & \cos(\phi) \end{bmatrix} \tag{3}$$

$$R_z(\psi) = \begin{bmatrix} \cos(\psi) & -\operatorname{sen}(\psi) & 0 \\ \operatorname{sen}(\psi) & \cos(\psi) & 0 \\ 0 & 0 & 1 \end{bmatrix} \tag{4}$$

$$R_y(\theta) = \begin{bmatrix} \cos(\theta) & 0 & \operatorname{sen}(\theta) \\ 0 & 1 & 0 \\ -\operatorname{sen}(\theta) & 0 & \cos(\theta) \end{bmatrix} \tag{5}$$

So that the inertial coordinates of position and the reference coordinates of the body are related by the three-dimensional rotation matrix, which is the equivalent of the multiplication of the rotation matrices of each axis as in (6) and (7).

where the matrix shown in (7) describes the rotation experienced by the reference frame of the body mirrored in the inertial reference frame, as shown in Figure 10.

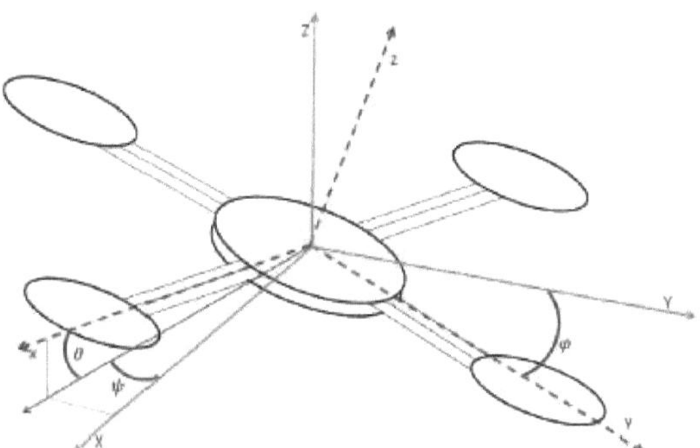

Figura 10 Rotación según los ángulos de Euler.
Figure 10 Rotation according to Euler angles.

As indicated in (7), this matrix provides the conversion from the quadcopter to the navigation frame. However, it may also be necessary in some cases to perform the transformation from the navigation frame to the vehicle frame, which can be achieved without major problem by performing the transposition of the rotation matrix defined in (7) as indicated in (8) and (9).

$$R_E^B = \left[R_B^E \right]^T \tag{8}$$

$$R_E^B = \begin{bmatrix} \cos(\theta)\cos(\psi) & \cos(\theta)\text{sen}(\psi) & -\text{sen}(\theta) \\ \text{sen}(\phi)\text{sen}(\theta)\cos(\psi) - \cos(\phi)\text{sen}(\psi) & \cos(\phi)\cos(\psi) + \text{sen}(\phi)\text{sen}(\theta)\text{sen}(\psi) & \text{sen}(\phi)\cos(\theta) \\ \text{sen}(\phi)\text{sen}(\psi) + \cos(\phi)\text{sen}(\theta)\cos(\psi) & \cos(\phi)\text{sen}(\theta)\text{sen}(\psi) - \text{sen}(\phi)\cos(\psi) & \cos(\phi)\cos(\theta) \end{bmatrix} \tag{9}$$

2.2.5. Modern control teona.

The modern trend in engineering systems is towards greater complexity, mainly because more complex tasks and good accuracy are required [55]. Complex systems can have multiple inputs and multiple outputs and can be time-varying.

Due to the need to meet increasingly demanding requirements on the behaviour of control systems, the increase in system complexity, and the easy access to programs and computers with high computational power, apart from data acquisition systems, modern control theory, which is a new approach to the analysis and design of complex control systems, has had development and applicability since the 1960's. This new approach is based on the concept of state, which as such is not new, since it has existed for quite some time in the area of classical dynamics and other fields. This new approach is based on the concept of state, which as such and in itself is not new, since it has existed for quite some time in the area of classical dynamics and other fields.

2.2.5.1.   Modern control theory versus classical control theory.

Modern control theory contrasts with conventional control theory in that its formulation is applicable to multi-input, multi-output systems, which may be linear or non-linear, time-invariant or time-varying, whereas classical control theory is only applicable to time-invariant, single-input, single-output systems. Moreover, modern control theory is essentially a time domain approximation, while classical control theory is a complex frequency domain approximation. Taking into account that a Linear Quadratic Regulator (LQR) is proposed for the control of the quadcopter, which belongs to the domain of modern control theory, it is necessary to define concepts such as: state, state variables, state vector, and state space, which are described below.

2.2.5.2.   Status.

The state of a dynamic system is the smallest set of variables (called *state variables*), so that knowledge of these variables at $t = t_0$, together with knowledge of the input for $t\$t_0$, completely determines the behaviour of the system at any $t\$t_0$.

Note that the concept of the state is not limited to risk systems. It is applicable to biological systems, economic systems, social systems and many others.

2.2.5.3.   State variables.

The variables of a dynamical system are the variables that constitute the smallest set of variables that determine the state of the dynamical system. If at least n variables x_1 , x_2 , ..., x_n are needed to completely describe the behaviour of a dynamical system (so that once the input for $t\$t_0$ is given and the initial state at $t = t_0$ is specified, the future state of the system is completely determined), then such n variables are a set of state variables.

Note that state variables need not be laughably measurable or observable quantities. State variables that do not represent laughable quantities and those that are neither measurable nor observable can be selected as state variables. Such freedom in the choice of state variables is an advantage of state space methods. However, from a practical point of view it is desirable to select for the state variables laughably measurable quantities, if possible, because the optimal control laws will require to feed back all state variables with an appropriate weighting.

2.2.5.4. State vector.

If *n* state variables are needed to fully describe the behaviour of a given system, then these *n* state variables can be considered as the *n* components of a vector **x**. This vector is called a state vector. This vector is called a *state vector*. A state vector is, therefore, a vector that uniquely determines the state of the system **x(t)** at any instant of time tt_0 , once the state at t = t_0 is known and the input *u(t)* is specified for tt_0 .

2.2.5.5. State space.

The n-dimensional space whose coordinate axes are formed by the *x1-axis*, *x-axis$_2$* , ..., *x-axis$_n$* , where x_1 , x_2 , ..., x_n are the state variables, is called state space. Any state can be represented as a point in the state space.

2.2.5.6. Equations in state space.

State space analysis focuses attention on the three types of variables that appear in the modelling of dynamic systems; input variables, output variables and state variables. Where the state space representation of a given system is not unique, unless the number of state variables is the same for any given state variable representation of any given system.

The dynamic system must contain elements that remember the input values for tt_1 . Since the integrators in a continuous time control system serve as a memory device, the outputs of such integrators can be considered as the variables describing the internal state of the dynamic system. Thus the outputs of integrators serve as state variables. The number of state variables to fully define the dynamics of the system is equal to the number of integrators appearing in the system.

Let be a multi-input multi-output system with *n* integrators. Suppose also that there are *r* inputs *u1(t)*, u_2 *(t)*, ..., u_r *(t)* and *m* outputs *y1(t)*, *y2(t)*, ..., *ym(t)*. The *n* outputs of the integrators are defined as state variables: x_1 *(t)*, x_2 *(t)*, ..., x_n *(t)*. The system can then be described by the set of equations given in (10) for the inputs, and in (11) for the outputs.

$$\dot{x}_1(t) = f_1\left(x_1, x_2, ..., x_n; u_1, u_2, ..., u_r; t\right)$$
$$\dot{x}_2(t) = f_2\left(x_1, x_2, ..., x_n; u_1, u_2, ..., u_r; t\right)$$
$$\vdots$$
$$\dot{x}_n(t) = f_n\left(x_1, x_2, ..., x_n; u_1, u_2, ..., u_r; t\right) \tag{10}$$

As for the outputs $y1(t), y2(t), ..., ym(t)$ of the system, they are obtained by the set of equations given in (11).

$$y_1(t) = g_1(x_1, x_2, ..., x_n; u_1, u_2, ..., u_r; t)$$
$$y_2(t) = g_2(x_1, x_2, ..., x_n; u_1, u_2, ..., u_r; t)$$
$$\vdots$$
$$y_m(t) = g_n(x_1, x_2, ..., x_n; u_1, u_2, ..., u_r; t)$$

(11)

If we now express the sets of equations in (10) and (11) as vectors and matrices, we have the system shown in (12), (13), and (14).

$$\dot{x}(t) = \begin{bmatrix} \dot{x}_1(t) \\ \dot{x}_2(t) \\ \vdots \\ \dot{x}_n(t) \end{bmatrix}, \quad y(t) = \begin{bmatrix} y_1(t) \\ y_2(t) \\ \vdots \\ y_m(t) \end{bmatrix}, \quad u(t) = \begin{bmatrix} u_1(t) \\ u_2(t) \\ \vdots \\ u_r(t) \end{bmatrix}$$

(12)

$$f(x, u, t) = \begin{bmatrix} f_1(x_1, x_2, ..., x_n; u_1, u_2, ..., u_r; t) \\ f_2(x_1, x_2, ..., x_n; u_1, u_2, ..., u_r; t) \\ \vdots \\ f_n(x_1, x_2, ..., x_n; u_1, u_2, ..., u_r; t) \end{bmatrix}$$

(13)

$$g(x, u, t) = \begin{bmatrix} g_1(x_1, x_2, ..., x_n; u_1, u_2, ..., u_r; t) \\ g_2(x_1, x_2, ..., x_n; u_1, u_2, ..., u_r; t) \\ \vdots \\ g_m(x_1, x_2, ..., x_n; u_1, u_2, ..., u_r; t) \end{bmatrix}$$

(14)

Which for simplicity can be expressed in equations (15) and (16).

$$\dot{x}(t) = f(x, u, t)$$

(15)

Where equation (15) is the equation of state and equation (16) is the equation of state.

$$y(t) = g(x, u, t)$$

(16)

of the output. If the vector functions **f** and/or **g** involve expHcitly time t, the system is called: Time-varying system.

If the system is known to deviate little or very little from the operating point, equations (15) and (16) can be linearised, yielding the equation of state (17) and the output equation (18).

$$\dot{\mathbf{x}}(t) = \mathbf{A}(t)\mathbf{x}(t) + \mathbf{B}(t)\mathbf{u}(t) \tag{17}$$

$$\mathbf{y}(t) = \mathbf{C}(t)\mathbf{x}(t) + \mathbf{D}(t)\mathbf{u}(t) \tag{18}$$

Where $\mathbf{A}(t)$ is called the state matrix, $\mathbf{B}(t)$ the input matrix, $\mathbf{C}(t)$ the output matrix, and $\mathbf{D}(t)$ the direct transmission matrix. Figure 11 shows a block diagram plotting the relationships of equations (17) and (18).

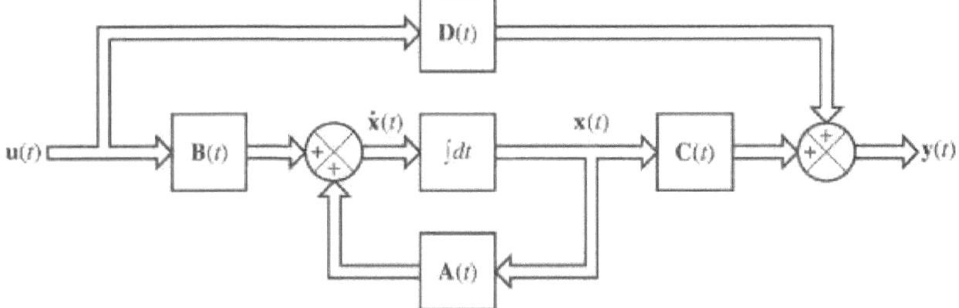

Figure 11 Block diagram of the linearised state-space system

If the vector functions $\mathbf{f}$ and $\mathbf{g}$ do not involve the time variable, t, explicitly, the system is called time invariant, whereby equations (17) and (18) are simplified as indicated in equations (19) and (20).

$$\dot{\mathbf{x}}(t) = \mathbf{A}\mathbf{x}(t) + \mathbf{B}\mathbf{u}(t) \tag{19}$$

$$\mathbf{y}(t) = \mathbf{C}\mathbf{x}(t) + \mathbf{D}\mathbf{u}(t) \tag{20}$$

2.2.6. Introduction to Optimal Control.

In modern control theory, the design method combining an observer and state feedback has been a fundamental tool in the control of state variable systems. However, it is not always the most useful method because of the following three difficulties [56]:

1. The translation of design specifications into pole locations is not straightforward, especially in complex systems; ^What is the best pole configuration for given specifications?

2. Feedback gains in MIMO systems are not unique; ^What is the best gain for a given pole configuration?

3. The observer's eigenvalues should be chosen faster than the controller's, but what additional criteria should be used to prefer one configuration over another?

The answers to the above questions are the ones that involve the development of Optimal Control, which is a mathematical technique that uses the state variables describing the process to solve optimisation problems in time-varying systems that are susceptible to being influenced by external forces. These time-varying systems are so diverse that they range from the human body, through economic systems, to planetary systems.

As such, the technique is based on choosing the state and observer feedback gains in a way that minimises a given optimisation criterion. Once the problem has been solved, optimal control gives us a behavioural path for the control variables, i.e., it indicates which actions must be followed in order to bring the whole system from an initial to a final state in an optimal way.

The particular optimisation criterion, or cost function, is a quadratic functional of the state and control input as indicated in equation (21).

$$J\big(x(t),u(t)\big) = \int_{t}^{T} \left[x^{T}(\tau)\,Q\,x(\tau) + u^{T}(\tau)\,R\,u(\tau) \right] d\tau \tag{21}$$

where Q and R are constant (but not necessarily) positive semidefinite and definite positive definite matrices respectively.

The control obtained by minimising this criterion is linear. As the criterion is based on quadratic functionals, the method is known as linear-quadratic (LQ: Linear-Quadratic), from which the linear-quadratic regulator (LQR: Linear-Quadratic-Regulator) is obtained.

Similar optimisation criteria are followed for the observer design, only that the functional will depend in this case on the estimation error, and is based on a statistical characterisation of the noise affecting the system, generally white noise or Gaussian noise due to the statistical characterisation of the noise used, where this optimal linear-quadratic estimator is more commonly known as the Kalman filter. When the LQ state feedback gain is combined with the Kalman filter, we obtain what is known as a Gaussian Linear Quadratic Controller (LQR).

2.2.6.1. The principle of optimality.

To facilitate the understanding of the state-variable optimisation criterion, a very simple form of graph theory will be used in which the nodes are the state of the system and the trajectories connecting them are the transitions from one state to another under the effect of an input to the system. See Figure 12.

When an optimisation criterion is associated with the system, each transition between states has a cost or penalty associated with it. For example, the cost may well be the distance between states such that, from a practical point of view, transitions that are too far away from the desired final state, or that take too long a path, or control actions of too high a value, may be penalised. As the system evolves from state to state, the costs add up to a total cost associated with the trajectory from the source state to the final state.

This concept can best be appreciated with the diagram in Figure 12, which represents 8 states of a system with their possible transitions. The initial state is 1, and the final state is 8. The system transitions from one state to another at each time *k* determined by the input *u[k]* and the equations *x[k+1]=Ax[k]+Bu[k]*, which is a discrete form of equation (19).

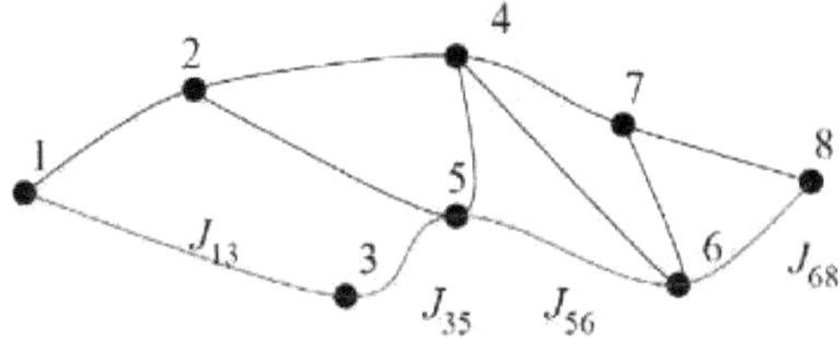

Figure 12 Possible trajectories from state 1 to 8.

The possible transitions are represented by the arcs connecting the initial state to the final state through the intermediate states. The cost associated with each transition is represented by the letter J; for example, the cost of moving from state 3 to state 5 is J_{35} .

Assuming that costs accumulate additively, we see that the path marked in red, for example, has a total cost $J_{13} + J_{35} + J_{56} + J_{68}$.

As there are several alternative paths from state 1 to state 8, the total cost will depend on the chosen path. The control signal *u[k]* that determines the least cost path is the *optimal strategy*. For continuous-time systems, the accumulation of costs is represented by integration, rather than by summation as in discrete systems.

It is fundamental to establish that the mathematical tool used to determine the optimal strategy is: *Bellman's optimality principle*, which is stated as follows:

At any intermediate point xi on an optimal path between x_0 and x_f, the strategy from xi to the end point xf must itself be an optimal strategy.

Although it may seem very obvious, this principle allows to solve in closed form the optimal control problems, apart from the fact that in practice it is also used the recursive computation of the optimal solutions in a procedure called: *dynamic programming*.

2.2.6.2. Continuous time LQ control.

The derivation of the LQ-optimal control for the continuous-time system starts with equation (19), where the "transitions" are reduced to infinitesimal "increments", so that the cost equation now manifests itself as shown in equation (22).

$$J\left(x(t_0),u(t_0),t_0\right)=\frac{1}{2}x^T\left(t_f\right)S\,x\left(t_f\right)+\frac{1}{2}\int_{t_0}^{t_f}\left(x^T(t)Q\,x(t)+u\,(t)R\,u(t)\right)dt \quad (22)$$

If we consider that the increments tend to zero, we obtain equations (23), (24), and (25) which define the optimal LQ control as indicated by Bay [57].

$$u^*(t)=-K(t)x(t) \quad (23)$$

$$K(t)=R^{-1}B^T P(t)x(t) \quad (24)$$

$$\dot{P}(t)=P(t)B\,R^{-1}B^T P(t)-Q-P(t)A-A^T P(t) \quad (25)$$

Equation (25) is the Riccati matrix differential equation, which must be solved backwards in time with "initial" conditions. $P\left(t_f\right)=S$ and the following considerations:

1. The optimal LQ control is a linear feedback of states, albeit unstable.

2. The "transients" in $P(t)$ and $K(t)$ will occur at the end of the interval. $\left[t_0,\,t_f\right]$.

In any case, the differential equation (25) is in general difficult to solve, even by numerical methods since: ^How can the almost infinite values of P(t) computed in inverse time be stored and then applied to the system. This difficulty leads to the search for a stationary optimal solution, $K(t) = K$, which arises from considering the *infinite horizon* case: $t_f \rightarrow \infty$.

Finally, it is pointed out that the optimality condition does not automatically guarantee good robustness properties [58]. In this sense, the stability of the closed-loop system must be reviewed separately after the design and implementation of the LQR controller, since trying to promote robustness in some of the system parameters may imply stochastic rather than deterministic conditions. The above condition can be interpreted to mean that the difference between two controllers in which only one of them is optimal is due to only some of the controller parameters being different [59].

3.1. Mathematical modelling of the system.

For the mathematical modelling that describes the behaviour of the quadrotor system, some of the equations of the doctoral thesis of Prof. Samir Bouabdallah [22] will be used. Selection based on the high degree of impact it has as it is one of the most referenced works in the various research articles published, and which were reviewed during the documentary research stage of this work.

3.2. Concepts and generalities.

The concept followed in this work for the dynamic modelling was to write the risic equations of the system, obtain the relevant parameters for the development of the computational model and identify only the dynamics of the actuators (propellers), which are the most important for the case of the quadrotor.

With this approach, the computational determination and construction of dynamic models of unsteady systems is made somewhat easier, since it is not possible to perform closed-loop parameter identification during the flight of the vehicle. The formalism of the Euler-Lagrange equations and the equations of a commercial DC motor is used to model with the highest plausible accuracy a possible working model. The formalism of the Newton-Euler equations, model identification, the aerodynamic elements of commercial rotor blades, and the theory of moments are used to model a quadrotor that is feasible to build. In any case, the model to be developed has the following considerations:

< The structure or body of the device is assumed to be rigid.

< The body of the structure has longitudinal and transverse symmetry.

< There is coincidence between the centre of gravity and the origin of the axes of symmetry.

< The propellers are rigid.

< The thrust and drag of the propellers are proportional to the square of the angular velocity of the latter.

The aerial devices under the principle of helicopter-type support are highly complex mechanical systems that combine numerous physical effects resulting from aerodynamics and the domain of mechanics [60]. One of the main effects to be considered in the modelling of the quadrotor is the gyroscopic effect, both for the rotation of the propellers and for the rotation of the assembly during flight manoeuvres. A short list of the main effects to which a quadcopter is subjected as established by Mullhaupt [61] is given in Table 4.1 below.

Table 4.1. Main effects acting on a quadcopter.

Effect	Source or Cause	Mathematical Formulation
Aerodynamics	Propeller rotation	$C \cdot \Omega^2$
	Blade shake-up.	

	Change in the rotational speed of the propeller.	$J \cdot \dot{\Omega}$
Inertial counter torque		
Gravity acceleration	Position of the centre of mass.	
Gyroscopic effects	Change in the orientation of the vehicle body	$I \cdot \theta \cdot \psi$
	Change in propeller plane orientation	$J \cdot \Omega, \cdot \theta, \phi$
Friction	With the air in all directions of movement	$C \cdot \dot{\phi}, C \cdot \dot{\theta}, C \cdot \dot{\psi}$

3.3. Rotation matrix, aeronautical consideration.

As indicated in section 2.2.4, the rotation of a rigid body in space can be parametrised using Euler angles. However, there are other methods such as: Quaternions or Tait-Bryan angles [62] more suitable for application in a real aircraft.

In particular, Tait-Bryan angles (also called Cardano angles) are widely used in Aerospace Engineering, where they can also be identified as adjusted Euler angles, since they are limited to the values at which the aircraft can rotate or partially rotate considering that the "absolute" horizontal position corresponds to an angular value of 0 in each rotation axis.

Particularly for aerospace engineering, the axes are referenced for an aircraft moving in the positive direction of the x-axis, i.e. the horizontal displacement to the left is considered positive so that the right side in the aircraft's forward movement corresponds to the positive direction of the y-axis, whereby the positive z-axis is downward and as already established in section 2.2.2, these angles correspond to pitch, lift, and yaw.

In particular, considering a coordinate system oriented according to the right hand rule, the three simplest rotations that can be made individually   and separately can be represented by:

< $R(x, \quad \phi)$, for rotation about the x-axis.

< $R(y, \quad o)$, for rotation about the y-axis.

< $R(z, \quad y)$, for rotation about the z-axis.

whose matrix expressions correspond to (3), (4), and (5), with the following limitations on the spin value: $\phi \in \left] -\frac{\pi}{2}, \frac{\pi}{2} \right[, \; \theta \in \left] -\frac{\pi}{2}, \frac{\pi}{2} \right[, \; \psi \in \left] -\frac{\pi}{2}, \frac{\pi}{2} \right[,$ which are established by the fact that the angular variations (derived with respect to time) in the Tait-Bryan angles can give rise to discontinuous functions, which is a very different condition from the angular variations of an aircraft (p, q, r) that are measured risically by gyroscopes or inertial measurement units (IMU) and commonly used in aerospace engineering to measure aircraft rotations and spatial orientation.

As for the mathematical aspect, the transformation matrix to take from $[p,q,r]^T$ a $[\dot{\phi},\dot{\theta},\dot{\psi}]^T$ is given according to Etkin [63] as follows:

$$\begin{bmatrix} p \\ q \\ r \end{bmatrix} = R_r \begin{bmatrix} \dot{\phi} \\ \dot{\theta} \\ \dot{\psi} \end{bmatrix} \tag{26}$$

the form of the transformation matrix being as follows:

3.4. Mathematical modelling of the quadcopter body.

$$R_r = \begin{bmatrix} 1 & 0 & -\operatorname{sen}\theta \\ 0 & \cos\phi & \operatorname{sen}\phi\cos\theta \\ 0 & -\operatorname{sen}\phi & \cos\phi\cos\theta \end{bmatrix} \tag{27}$$

For the determination of the dynamic model of any aircraft, use is made of the Euler-Lagrange set of considerations [2] and the assumptions set out in section 4.2, recalling that the Lagrangian = Kinetic Energy - Potential Energy, i.e:

$$L = T - V \tag{28}$$

and the general form of the equations of motion by the Lagrangian method is:

$$\Gamma_i = \frac{d}{dt}\left(\frac{\delta L}{\delta \dot{q}_i}\right) - \frac{\delta L}{\delta q_i} \tag{29}$$

where q_i is the generalised coordinate system, and Γ is the generalised force system.

As set out in Figure 8, the inertial coordinate system set at the earth, E, with the orthogonal base defined by: $R [\vec{X},\vec{Y},\vec{Z}]$, and the moving system of

coordinates fixed in the vehicle. B, with the orthogonal basis defined by: $[\vec{x},\vec{y},\vec{z}]$ and as previously stated, if any point of the vehicle undergoes three successive rotations, it is possible to express its spatial location at any time using the expression:

$$r_{X,Y,Z}(x,y,z) = R(\phi,\theta,\psi)\begin{pmatrix} x \\ y \\ z \end{pmatrix} \tag{30}$$

$$r_X(x,y,z) = x(\cos\psi\cos\theta) + y(\cos\psi\,\mathrm{sen}\,\theta\,\mathrm{sen}\,\phi - \mathrm{sen}\,\psi\cos\phi) + z(\cos\psi\,\mathrm{sen}\,\theta\cos\phi + \mathrm{sen}\,\psi\,\mathrm{sen}\,\phi)$$

$$r_Y(x,y,z) = x(\mathrm{sen}\,\psi\cos\theta) + y(\mathrm{sen}\,\psi\,\mathrm{sen}\,\theta\,\mathrm{sen}\,\phi - \cos\psi\cos\phi) + z(\mathrm{sen}\,\psi\,\mathrm{sen}\,\theta\cos\phi - \cos\psi\,\mathrm{sen}\,\phi) \tag{31}$$

$$r_Z(x,y,z) = x(-\mathrm{sen}\,\theta) \qquad + y(\mathrm{sen}\,\phi\cos\phi) \qquad\qquad + z(\cos\phi\cos\theta)$$

This expression, when derived with respect to time, gives the corresponding velocities in each axis:

$$
\begin{aligned}
v_X(x,y,z) = \; & x\left(-\mathrm{sen}\,\theta\cos\psi\,\dot\theta - \cos\theta\,\mathrm{sen}\,\psi\,\dot\psi\right) \\
+ & y\left(-\cos\psi\cos\phi\,\dot\psi + \mathrm{sen}\,\psi\,\mathrm{sen}\,\phi\,\dot\phi - \mathrm{sen}\,\psi\,\mathrm{sen}\,\phi\,\mathrm{sen}\,\theta\,\dot\psi + \cos\psi\cos\phi\,\mathrm{sen}\,\theta\,\dot\phi + \cos\psi\,\mathrm{sen}\,\phi\cos\theta\,\dot\theta\right) \\
+ & z\left(\cos\psi\,\mathrm{sen}\,\phi\,\dot\psi + \mathrm{sen}\,\psi\cos\phi\,\dot\phi - \mathrm{sen}\,\psi\cos\phi\,\mathrm{sen}\,\theta\,\dot\psi - \cos\psi\,\mathrm{sen}\,\phi\,\mathrm{sen}\,\theta\,\dot\phi + \cos\psi\cos\phi\cos\theta\,\dot\theta\right)
\end{aligned} \tag{32}
$$

$$
\begin{aligned}
v_Y(x,y,z) = \; & x\left(-\mathrm{sen}\,\theta\,\mathrm{sen}\,\psi\,\dot\theta + \cos\theta\cos\psi\,\dot\psi\right) \\
+ & y\left(-\mathrm{sen}\,\psi\cos\phi\,\dot\psi - \cos\psi\,\mathrm{sen}\,\phi\,\dot\phi + \cos\psi\,\mathrm{sen}\,\phi\,\mathrm{sen}\,\theta\,\dot\psi + \mathrm{sen}\,\psi\cos\phi\,\mathrm{sen}\,\theta\,\dot\phi + \mathrm{sen}\,\psi\,\mathrm{sen}\,\phi\cos\theta\,\dot\theta\right) \\
+ & z\left(\mathrm{sen}\,\psi\,\mathrm{sen}\,\phi\,\dot\psi + \cos\psi\cos\phi\,\dot\phi + \cos\psi\cos\phi\,\mathrm{sen}\,\theta\,\dot\psi - \mathrm{sen}\,\psi\,\mathrm{sen}\,\phi\,\mathrm{sen}\,\theta\,\dot\phi + \mathrm{sen}\,\psi\cos\phi\cos\theta\,\dot\theta\right)
\end{aligned} \tag{33}
$$

$$
\begin{aligned}
v_Z(x,y,z) = \; & x\left(-\cos\theta\,\dot\theta\right) \\
+ & y\left(\cos\phi\cos\theta\,\dot\phi - \mathrm{sen}\,\phi\,\mathrm{sen}\,\theta\,\dot\theta\right) \\
+ & z\left(-\mathrm{sen}\,\phi\cos\theta\,\dot\phi - \cos\phi\,\mathrm{sen}\,\theta\,\dot\theta\right)
\end{aligned} \tag{34}
$$

which for simplicity can be expressed as:

$$v_X(x,y,z) = x v_{Xx} + y v_{Xy} + z v_{Xz} = \begin{pmatrix} v_{Xx} & v_{Xy} & v_{Xz} \end{pmatrix}\begin{pmatrix} x \\ y \\ z \end{pmatrix} \tag{35}$$

$$v_Y(x,y,z) = x v_{Yx} + y v_{Yy} + z v_{Yz} = \begin{pmatrix} v_{Yx} & v_{Yy} & v_{Yz} \end{pmatrix}\begin{pmatrix} x \\ y \\ z \end{pmatrix} \tag{36}$$

$$v_Z(x,y,z) = x v_{Zx} + y v_{Zy} + z v_{Zz} = \begin{pmatrix} v_{Zx} & v_{Zy} & v_{Zz} \end{pmatrix}\begin{pmatrix} x \\ y \\ z \end{pmatrix} \tag{37}$$

Where the quadratic magnitude of the velocity for any point is given by:

$$v^2(x,y,z) = v_X^2(x,y,z) + v_Y^2(x,y,z) + v_Z^2(x,y,z) \tag{38}$$

$$v^2(x,y,z) = \begin{pmatrix} v_{Xx} & v_{Xy} & v_{Xz} \end{pmatrix} \Lambda \begin{pmatrix} v_{Xx} \\ v_{Xy} \\ v_{Xz} \end{pmatrix} + \begin{pmatrix} v_{Yx} & v_{Yy} & v_{Yz} \end{pmatrix} \Lambda \begin{pmatrix} v_{Yx} \\ v_{Yy} \\ v_{Yz} \end{pmatrix} + \begin{pmatrix} v_{Zx} & v_{Zy} & v_{Zz} \end{pmatrix} \Lambda \begin{pmatrix} v_{Zx} \\ v_{Zy} \\ v_{Zz} \end{pmatrix} \tag{39}$$

Con:

$$\Lambda = \begin{pmatrix} x^2 & xy & xz \\ xy & y^2 & yz \\ xz & yz & z^2 \end{pmatrix} \tag{40}$$

And where:

$$
\begin{aligned}
v^2(x,y,z) =\ & x^2 \cdot \left(v_{Xx}^2 + v_{Yx}^2 + v_{Zx}^2 \right) \\
& + y^2 \cdot \left(v_{Xy}^2 + v_{Yy}^2 + v_{Zy}^2 \right) \\
& + z^2 \cdot \left(v_{Xz}^2 + v_{Yz}^2 + v_{Zz}^2 \right) \\
& + 2xy \cdot \left(v_{Xx} \cdot v_{Xy} + v_{Yx} \cdot v_{Yy} + v_{Zx} \cdot v_{Zy} \right) \\
& + 2xz \cdot \left(v_{Xx} \cdot v_{Xz} + v_{Yx} \cdot v_{Yz} + v_{Zx} \cdot v_{Zz} \right) \\
& + 2yz \cdot \left(v_{Xy} \cdot v_{Xz} + v_{Yy} \cdot v_{Yz} + v_{Zy} \cdot v_{Zz} \right)
\end{aligned}
\tag{41}
$$

$$
\begin{aligned}
v^2(x,y,z) =\ & x^2\left(\cos^2\theta\,\dot{\psi}^2 + \dot{\theta}^2\right) \\
& + y^2\left(\dot{\psi}^2\left(\cos^2\phi + \mathrm{sen}^2\phi\,\mathrm{sen}^2\theta\right) + \dot{\psi}\left(-2\,\mathrm{sen}\phi\cos\phi\cos\theta\,\dot{\theta} - 2\dot{\phi}\,\mathrm{sen}\theta\right) + \mathrm{sen}^2\phi\,\dot{\theta}^2 + \dot{\phi}^2\right) \\
& + z^2\left(\dot{\psi}^2\left(\mathrm{sen}^2\phi + \cos^2\phi\,\mathrm{sen}^2\theta\right) + \dot{\psi}\left(2\,\mathrm{sen}\phi\cos\phi\cos\theta\,\dot{\theta} - 2\dot{\phi}\,\mathrm{sen}\theta\right) + \cos^2\phi\,\dot{\theta}^2 + \dot{\phi}^2\right) \\
& + 2xy\left(\dot{\psi}^2\,\mathrm{sen}\phi\,\mathrm{sen}\theta\cos\theta + \dot{\psi}\left(\cos\phi\,\mathrm{sen}\theta\,\dot{\theta} - \mathrm{sen}\phi\cos\theta\,\dot{\phi}\right) - \cos\phi\,\dot{\phi}\dot{\theta}\right) \\
& + 2xz\left(\dot{\psi}^2\cos\phi\,\mathrm{sen}\theta\cos\theta + \dot{\psi}\left(-\cos\phi\cos\theta\,\dot{\phi} - \mathrm{sen}\phi\,\mathrm{sen}\theta\,\dot{\theta}\right) + \mathrm{sen}\phi\,\dot{\phi}\dot{\theta}\right) \\
& + 2yz\left(-\dot{\psi}^2\,\mathrm{sen}\phi\cos\phi\cos^2\theta + \dot{\psi}\left(\mathrm{sen}^2\phi\cos\theta\,\dot{\theta} - \cos^2\phi\cos\theta\,\dot{\theta}\right) + \mathrm{sen}\phi\cos\phi\,\dot{\theta}^2\right)
\end{aligned}
\tag{42}
$$

it is possible to write the above expression as:

$$
\begin{aligned}
v^2(x,y,z) =\ & \left(y^2 + z^2\right)\left(\dot{\psi}^2\,\mathrm{sen}^2\theta - 2\,\mathrm{sen}\theta\,\dot{\phi}\dot{\psi} + \dot{\phi}^2\right) \\
& + \left(x^2 + z^2\right)\left(\dot{\psi}^2\,\mathrm{sen}^2\phi\cos^2\theta + 2\,\mathrm{sen}\phi\cos\phi\cos\theta\,\dot{\theta}\dot{\psi} + \cos^2\phi\,\dot{\theta}^2\right) \\
& + \left(x^2 + y^2\right)\left(\dot{\psi}^2\cos^2\phi\cos^2\theta - 2\,\mathrm{sen}\phi\cos\phi\cos\theta\,\dot{\theta}\dot{\psi} + \mathrm{sen}^2\phi\,\dot{\theta}^2\right) \\
& + 2xy\left(\dot{\psi}^2\,\mathrm{sen}\phi\,\mathrm{sen}\theta\cos\theta + \dot{\psi}\left(\cos\phi\,\mathrm{sen}\theta\,\dot{\theta} - \mathrm{sen}\phi\cos\theta\,\dot{\phi}\right) - \cos\phi\,\dot{\phi}\dot{\theta}\right) \\
& + 2xz\left(\dot{\psi}^2\cos\phi\,\mathrm{sen}\theta\cos\theta + \dot{\psi}\left(-\cos\phi\cos\theta\,\dot{\phi} - \mathrm{sen}\phi\,\mathrm{sen}\theta\,\dot{\theta}\right) + \mathrm{sen}\phi\,\dot{\phi}\dot{\theta}\right) \\
& + 2yz\left(-\dot{\psi}^2\,\mathrm{sen}\phi\cos\phi\cos^2\theta + \dot{\psi}\left(\mathrm{sen}^2\phi\cos\theta\,\dot{\theta} - \cos^2\phi\cos\theta\,\dot{\theta}\right) + \mathrm{sen}\phi\cos\phi\,\dot{\theta}^2\right)
\end{aligned}
\tag{43}
$$

As for the expression for the kinetic energy, we have the following equation:

$$T = \frac{1}{2}\int y^2 + z^2(R)dm(r)\cdot\left(\dot\phi^2 - \dot\psi\dot\phi 2\,\mathrm{sen}\theta + \dot\psi^2\,\mathrm{sen}^2\theta\right)$$

$$+ \frac{1}{2}\int z^2 + x^2(R)dm(r)\cdot\left(\dot\theta^2\cos^2\phi + \dot\theta\dot\psi 2\,\mathrm{sen}\phi\cos\phi\cos\theta + \dot\psi^2\,\mathrm{sen}^2\phi\cos^2\theta\right)$$

$$+ \frac{1}{2}\int x^2 + y^2(R)dm(r)\cdot\left(\dot\theta^2\,\mathrm{sen}^2\phi - \dot\theta\dot\psi 2\,\mathrm{sen}\phi\cos\phi\cos\theta + \dot\psi^2\cos^2\phi\cos^2\theta\right) \qquad (44)$$

$$+ \int xy(R)dm(r)\cdot\left(\dot\psi^2\,\mathrm{sen}\phi\,\mathrm{sen}\theta\cos\theta + \dot\psi\left(\cos\phi\,\mathrm{sen}\theta\dot\theta - \mathrm{sen}\phi\cos\theta\dot\phi\right) - \cos\phi\dot\phi\dot\theta\right)$$

$$+ \int xz(R)dm(r)\cdot\left(\dot\psi^2\cos\phi\,\mathrm{sen}\theta\cos\theta + \dot\psi\left(-\cos\phi\cos\theta\dot\phi - \mathrm{sen}\phi\,\mathrm{sen}\theta\dot\theta\right) + \mathrm{sen}\phi\dot\phi\dot\theta\right)$$

$$+ \int yz(R)dm(r)\cdot\left(-\dot\psi^2\,\mathrm{sen}\phi\cos\phi\cos^2\theta + \dot\psi\left(\mathrm{sen}^2\phi\cos\theta\dot\theta - \cos^2\phi\cos\theta\dot\theta\right) + \mathrm{sen}\phi\cos\phi\dot\theta^2\right)$$

where the moments of inertia (elements of the main diagonal of the matrix) and the products of inertia (elements bordering the main diagonal) appear.

One of the advantages of the mechanical symmetry of the quadrotor is that it allows the products of inertia to be ignored since the magnitude of these are generally a thousand times smaller than the moments of inertia so that the inertia matrix is simplified to just the main diagonal. With this consideration, the expression for the kinetic energy becomes:

$$T = \frac{1}{2}Ixx\left(\dot\phi - \dot\psi\,\mathrm{sen}\theta\right)^2 + \frac{1}{2}Iyy\left(\dot\theta\cos\phi + \dot\psi\,\mathrm{sen}\phi\cos\theta\right)^2 + \frac{1}{2}Izz\left(\dot\theta\,\mathrm{sen}\phi - \dot\psi\cos\phi\cos\theta\right)^2 \quad (45)$$

On the other hand, the expression for Potential Energy becomes:

$$V = g\int\left(-x\,\mathrm{sen}\theta + y\,\mathrm{sen}\phi\cos\theta + z\cos\phi\cos\theta\right)dm(r) \qquad (46)$$

which may well be expressed as follows:

$$V = \int x\,dm(x)\cdot\left(-g\,\mathrm{sen}\theta\right) + \int y\,dm(y)\cdot\left(g\,\mathrm{sen}\phi\cos\theta\right) + \int z\,dm(z)\cdot\left(g\cos\phi\cos\theta\right) \qquad (47)$$

substituting now in the lagrangian (29) and considering the equation of motion we have:

$$\frac{d}{dt}\left(\frac{\partial L}{\partial\dot\phi}\right) - \frac{\partial L}{\partial\phi} = \tau_\phi$$

$$\frac{d}{dt}\left(\frac{\partial L}{\partial\dot\theta}\right) - \frac{\partial L}{\partial\theta} = \tau_\theta \qquad (48)$$

$$\frac{d}{dt}\left(\frac{\partial L}{\partial\dot\psi}\right) - \frac{\partial L}{\partial\psi} = \tau_\psi$$

that after developing successively for each of the equations of: inclination, elevation, and turning, the following equations are obtained:

$$\frac{d}{dt}\left(\frac{\partial L}{\partial\dot\phi}\right) - \frac{\partial L}{\partial\phi} = \ddot\phi\cdot Ixx - \ddot\psi\cdot\mathrm{sen}\theta\cdot Ixx - \dot\psi\dot\theta\cdot\cos\theta\left(Ixx + (Iyy - Izz)(2\cos^2\phi - 1)\right)$$

$$+ \dot\theta^2\cdot\frac{1}{2}\,\mathrm{sen}2\phi(Iyy\text{-}Izz) - \dot\psi^2\cdot\frac{1}{2}\,\mathrm{sen}2\phi\cos\theta^2(Iyy - Izz) \qquad (49)$$

$$+ \int y\,dm(y)\cdot\left(-g\cos\phi\cos\theta\right) + \int z\,dm(z)\cdot\left(g\,\mathrm{sen}\phi\cos\theta\right)$$

$$\frac{d}{dt}\left(\frac{\partial L}{\partial \dot{\theta}}\right) - \frac{\partial L}{\partial \theta} = \ddot{\theta}\cdot\left(Iyy\cos^2\phi + Izz\operatorname{sen}^2\phi\right) + \ddot{\psi}\cdot\frac{1}{2}\operatorname{sen}2\phi\cos\theta\cdot\left(Iyy - Izz\right)$$

$$+\,\dot{\psi}^2\cdot\frac{1}{2}\operatorname{sen}2\theta\left(Ixx + Iyy\operatorname{sen}^2\phi + Izz\cos^2\phi\right) + \dot{\theta}\dot{\phi}\cdot\operatorname{sen}2\phi(Izz\text{-}Iyy)$$

$$+\,\dot{\psi}\dot{\phi}\cdot\cos\theta\left(\cos2\phi\cdot(Iyy - Izz) + Ixx\right) + \int xdm(x)\cdot(-g\cos\theta)$$

$$-\int ydm(y)\cdot(g\operatorname{sen}\phi\operatorname{sen}\theta) - \int zdm(z)\cdot(g\cos\phi\operatorname{sen}\theta) \tag{50}$$

$$\frac{d}{dt}\left(\frac{\partial L}{\partial \dot{\psi}}\right) - \frac{\partial L}{\partial \psi} = \ddot{\psi}\cdot\left(\cos^2\theta\left(Izz\cos^2\phi + Iyy\operatorname{sen}^2\phi\right) + \operatorname{sen}^2\theta Ixx\right) - \ddot{\phi}\cdot\operatorname{sen}\theta\, Ixx$$

$$+\,\dot{\theta}\cdot\frac{1}{2}\operatorname{sen}2\phi\cos\theta(Iyy - Izz) + \dot{\theta}\dot{\psi}\cdot\operatorname{sen}2\theta\left(Ixx - Izz\cos^2\phi + Iyy\operatorname{sen}^2\phi\right)$$

$$-\,\dot{\psi}\dot{\phi}\cdot\operatorname{sen}2\phi\cos^2\theta(Iyy - Izz) + \dot{\theta}\dot{\phi}\cdot\cos\theta\left(Ixx + (2\cos^2\phi - 1)(Iyy - Izz)\right)$$

$$-\,\dot{\theta}^2\cdot\frac{1}{2}\operatorname{sen}2\phi\operatorname{sen}\theta(Iyy - Izz) \tag{51}$$

These equations can be simplified by expressing the velocities and accelerations according to the Euler angles in the inertial reference system as a function of the instantaneous velocity and acceleration of the velocity according to the moving coordinate system by using the transformation matrix given in (27), so that the Lagrangians are expressed as follows:

$$\frac{d}{dt}\left(\frac{\partial L}{\partial \dot{\phi}}\right) - \frac{\partial L}{\partial \phi} = Ixx\,\dot{\omega}_x - (Iyy - Izz)\omega_y\omega_z$$

$$+\int ydm(y)\cdot(-g\cos\phi\cos\theta) + \int zdm(z)\cdot(+g\operatorname{sen}\phi\cos\theta) \tag{52}$$

$$\frac{d}{dt}\left(\frac{\partial L}{\partial \dot{\theta}}\right) - \frac{\partial L}{\partial \theta} = -\operatorname{sen}\phi\left(\dot{\omega}_z Izz - \omega_x\omega_y(Ixx - Iyy)\right) + \cos\phi\left(\dot{\omega}_y\cdot Iyy - \omega_x\omega_z(Izz - Ixx)\right)$$

$$+\int xdm(x)\cdot(-g\cos\theta) - \int ydm(y)\cdot(g\operatorname{sen}\phi\operatorname{sen}\theta) - \int zdm(z)\cdot(g\cos\phi\operatorname{sen}\theta) \tag{53}$$

$$\frac{d}{dt}\left(\frac{\partial L}{\partial \dot{\psi}}\right) - \frac{\partial L}{\partial \psi} = -\operatorname{sen}\theta\cdot\left(\dot{\omega}_x Ixx - \omega_y\omega_z(Iyy - Izz)\right) + \operatorname{sen}\phi\cos\theta\cdot\left(\dot{\omega}_y Iyy - \omega_x\omega_z(Izz - Ixx)\right)$$

$$+\cos\phi\cos\theta\cdot\left(\dot{\omega}_z Izz - \omega_x\omega_y(Ixx - Iyy)\right) \tag{54}$$

The non-conservative moments being as follows:

$bl\left(\Omega_4^2 - \Omega_2^2\right)$ It is the thrust imbalance between engines 4 and 2.

$bl\left(\Omega_3^2 - \Omega_1^2\right)$ It is the thrust imbalance between engines 3 and 1.

$d\left(\Omega_1^2 - \Omega_2^2 + \Omega_3^2 - \Omega_4^2\right)$ It is the imbalance in thrust between the propellers $(1, 3)$ and $(2, 4)$.

$J_r\omega_y\left(\Omega_1 + \Omega_3 - \Omega_2 - \Omega_4\right)$ It is the gyroscopic effect due to the rotation of the propellers.

$J_r\omega_x\left(-\Omega_1 - \Omega_3 + \Omega_2 + \Omega_4\right)$ It is the gyroscopic effect due to the rotation of the propellers.

where the total moment acting on the x, y, and z axes is:

$$\tau_x = bl\left(\Omega_4^2 - \Omega_2^2\right) + J_r\omega_y\left(\Omega_1 + \Omega_3 - \Omega_2 - \Omega_4\right) \tag{55}$$

$$\tau_y = bl\left(\Omega_3^2 - \Omega_1^2\right) + J_r\omega_x\left(-\Omega_1 - \Omega_3 + \Omega_2 + \Omega_4\right) \tag{56}$$

$$\tau_z = d\left(\Omega_1^2 - \Omega_2^2 + \Omega_3^2 - \Omega_4^2\right) \tag{57}$$

If we now consider a small deflection at all angles of the aircraft, the dynamics of the rotation subsystems becomes:

$$\ddot{\phi} = \frac{J_r\dot{\theta}\left(\Omega_1 + \Omega_3 - \Omega_2 - \Omega_4\right)}{Ixx} + \frac{Iyy - Izz}{Ixx}\dot{\psi}\dot{\theta} + \frac{bl\left(\Omega_2^2 - \Omega_4^2\right)}{Ixx} \tag{58}$$

$$\ddot{\theta} = \frac{J_r\dot{\phi}\left(-\Omega_1 - \Omega_3 + \Omega_2 + \Omega_4\right)}{Iyy} + \frac{Izz - Ixx}{Iyy}\dot{\psi}\dot{\phi} + \frac{bl\left(\Omega_3^2 - \Omega_1^2\right)}{Iyy} \tag{59}$$

$$\ddot{\psi} = \frac{d\left(\Omega_1^2 - \Omega_2^2 + \Omega_3^2 - \Omega_4^2\right)}{Izz} + \frac{Ixx - Iyy}{Izz}\dot{\theta}\dot{\phi} \tag{60}$$

3.5. Mathematical modelling of the engine and propeller.

For the modelling of the motor, it will be considered to be a standard DC motor which obeys the following equations:

$$L\frac{di(t)}{dt} = u - R_{mot}i(t) - k_e\omega_m \tag{61}$$

$$J_m\frac{d\omega_m}{dt} = M_{em} - M_{fr} \tag{62}$$

where $M_{em} = k_m \cdot i(t)$ is the torque of the motor and M_{fr} is the opposing moment due to friction caused by the drag of the propeller. Where, if the inductance, L, of the motor is ignored because of its small size, we have then that:

$$i(t) = \frac{u - k_e\omega_m}{R_{mot}} \tag{63}$$

where:

$$J_m\frac{d\omega_m}{dt} = k_m \cdot \frac{u - k_e\omega_m}{R_{mot}} - M_{fr} \tag{64}$$

which is equivalent to:

$$J_m \dot{\omega}_m = -\frac{k_m^2}{R_{mot}} \omega_m - M_{fr} + \frac{k_m}{R_{mot}} u \tag{65}$$

in the case where the propeller is coupled to the engine by means of a speed reduction gear, the opposing moment due to friction caused by aerodynamic drag is as follows:

$$M_{fr} = \frac{d\omega_m^2}{\eta r} \tag{66}$$

where n is the efficiency of the speed reduction gear and ε is the reduction ratio. Which is equivalent to:

$$\omega_{propela} = \frac{\omega_m}{r} \Rightarrow M_{fr} = \frac{d}{\eta r^3} \omega_m^2 \tag{67}$$

The inertia seen by the engine is:

$$J_{propela} \omega_{propela}^2 = \eta \cdot J_{popela \to motor} \omega_m^2 \Rightarrow J_{propela \to motor} = \frac{J_{propela}}{\eta r^2} \tag{68}$$

if the inertia of the propeller seen by the engine is $J_{propela \to motor}$. The engine equation can be expressed in the following form:

If J_t is the total inertia seen by the engine, then we have that:

equation which can be rewritten if it is considered that: $\dfrac{1}{\tau} = \dfrac{k_m^2}{R J_t}$

$$\dot{\omega}_m = -\frac{1}{\tau} \omega_m - \frac{d}{\eta r^3 J_t} \omega_m^2 + \frac{1}{k_m \tau} u \tag{71}$$

If it is considered that the rotational speed of the propeller, d_m, undergoes small variations around the operating speed, the above equation can be linearised around d_0 by means of a first order Taylor series, so that (7_) can be rewritten with the form:

$$\dot{\omega}_m = -A \omega_m + Bu + C$$
$$\text{Con: } A = \frac{1}{\tau} + \frac{2 d \omega_0}{\eta r^3 J_t}, \quad B = \frac{1}{k_m \tau}, \quad C = \frac{d \omega_0}{\eta r^3 J_t} \tag{72}$$

In this way, we obtain the expression that indicates the rotational speed required in the thrusters to develop the different flight manoeuvres necessary to undertake the flight path.

4.1. Control strategy.

In this section, the development of the linear quadratic controller with six degrees of freedom will be considered with the use of only four control actions, which correspond to the rotational speed of the rotors. However, before entering into the application of the optimal control theory, the set of equations developed in the previous section and corresponding to the mathematical modelling of the aircraft will be put in "order" in order to establish the variables that will define the state space of the aircraft and from these the referred controller will be derived.

4.2. Summary of the mathematical model of the quadrotor.

According to the mathematical development of the previous chapter, it can be established that the vector defined by $[x,y,z,\phi,\theta,\psi]^T$ is the vector containing the linear and angular position coordinates of the quadrotor in the inertial coordinate system, and the vector defined by $[u,v,w,p,q,r]^T$ is the vector containing the linear and angular position coordinates of the quadrotor in the moving coordinate system. From the equations on the dynamics of the vehicle in three dimensions, it can be established that the two reference systems are linked by the following relations:

$$v = R \cdot v_B \tag{73}$$
$$\omega = T \cdot \omega_B \tag{74}$$

where $v = [\dot{x},\dot{y},\dot{z}]^T \in R^3$, $\omega = [\dot{\phi},\dot{\theta},\dot{\psi}]^T \in R^3$, $v_B = [u,v,w]^T \in R^3$, $\omega_B = [p,q,r]^T \in R^3$, with R is the rotational matrix defined in (7), and T is the matrix for the angular transformations according to Lee [64].

$$T = \begin{bmatrix} 1 & \operatorname{sen}\phi\tan\theta & \cos\phi\tan\theta \\ 0 & \cos\phi & -\operatorname{sen}\phi \\ 0 & \dfrac{\operatorname{sen}\phi}{\cos\theta} & \dfrac{\cos\phi}{\cos\theta} \end{bmatrix} \tag{75}$$

so that the kinematic model of the quadrotor can be summarised as follows:

$$\dot{x} = w\left[\text{sen}\phi\ \text{sen}\psi + \cos\phi\cos\psi\ \text{sen}\theta\right] - v\left[\cos\phi\ \text{sen}\psi - \cos\psi\ \text{sen}\phi\ \text{sen}\theta\right] + u\left[\cos\psi\cos\theta\right]$$

$$\dot{y} = v\left[\cos\phi\cos\psi + \text{sen}\phi\ \text{sen}\psi\ \text{sen}\theta\right] - w\left[\cos\psi\ \text{sen}\phi - \cos\phi\ \text{sen}\psi\ \text{sen}\theta\right] + u\left[\cos\theta\ \text{sen}\psi\right]$$

$$\dot{z} = w\left[\cos\phi\cos\theta\right] - u\left[\text{sen}\theta\right] + v\left[\cos\theta\ \text{sen}\phi\right]$$

$$\dot{\phi} = p + r\left[\cos\phi\tan\theta\right] + q\left[\text{sen}\phi\tan\theta\right] \tag{76}$$

$$\dot{\theta} = q\left[\cos\phi\right] - r\left[\text{sen}\phi\right]$$

$$\dot{\psi} = r\frac{\cos\phi}{\cos\theta} + q\frac{\text{sen}\phi}{\cos\theta}$$

and the complete dynamic model, considering the forces, of the quadrotor in the moving coordinate system in which *f* represents the forces, *m* the mass of the vehicle, *I* the inertias of the system, *g* the acceleration of gravity, *m* the torques due to the effect of the rotors, τ_w the torques due to the effect of the wind gusts, becomes:

$$-mg\left[\text{sen}\theta\right] + f_{wx} = m\left(\dot{u} + qw - rv\right)$$

$$mg\left[\cos\theta\ \text{sen}\phi\right] + f_{wy} = m\left(\dot{v} - pw + ru\right)$$

$$mg\left[\cos\theta\cos\phi\right] + f_{wz} - f_t = m\left(\dot{w} + pv - qu\right)$$

$$\tau_x + \tau_{wx} = \dot{p}I_x - qrI_y + qrI_z \tag{77}$$

$$\tau_y + \tau_{wy} = \dot{q}I_y + prI_x - prI_z$$

$$\tau_z + \tau_{wz} = \dot{r}I_z - pqI_x + pqI_y$$

4.3. Model in state space.

As stated in the previous summary, the state space vector describing the kinematics and dynamics of the air vehicle is defined as follows:

$$X = \left[\phi, \theta, \psi, p, q, r, u, v, w, x, y, z\right]^T \in \mathbb{R}^{12} \tag{78}$$

and with this definition it is possible to rewrite the equations describing the complete quadrotor dynamics set out in (76) and (77) as follows:

$$\dot{\phi} = p + r[\cos\phi\tan\theta] + q[\operatorname{sen}\phi\tan\theta]$$

$$\dot{\theta} = q[\cos\phi] - r[\operatorname{sen}\phi]$$

$$\dot{\psi} = r\frac{\cos\phi}{\cos\theta} + q\frac{\operatorname{sen}\phi}{\cos\theta}$$

$$\dot{p} = \frac{I_y - I_z}{I_x}rq + \frac{\tau_x + \tau_{wx}}{I_x}$$

$$\dot{q} = \frac{I_z - I_x}{I_y}pr + \frac{\tau_y + \tau_{wy}}{I_y}$$

$$\dot{r} = \frac{I_x - I_y}{I_z}pq + \frac{\tau_z + \tau_{wz}}{I_z} \tag{79}$$

$$\dot{u} = rv - qw - g[\operatorname{sen}\theta] + \frac{f_{wx}}{m}$$

$$\dot{v} = pw - ru + g[\operatorname{sen}\phi\cos\theta] + \frac{f_{wy}}{m}$$

$$\dot{w} = qu - pv + g[\cos\theta\cos\phi] + \frac{f_{wz} - f_t}{m}$$

$$\dot{x} = w[\operatorname{sen}\phi\operatorname{sen}\psi + \cos\phi\cos\psi\operatorname{sen}\theta] - v[\cos\phi\operatorname{sen}\psi - \cos\psi\operatorname{sen}\phi\operatorname{sen}\theta] + u[\cos\psi\cos\theta]$$

$$\dot{y} = v[\cos\phi\cos\psi + \operatorname{sen}\phi\operatorname{sen}\psi\operatorname{sen}\theta] - w[\operatorname{sen}\phi\cos\psi - \operatorname{sen}\psi\cos\phi\operatorname{sen}\theta] + u[\operatorname{sen}\psi\cos\theta]$$

$$\dot{z} = w[\cos\phi\cos\theta] - u[\operatorname{sen}\theta] + v[\cos\theta\operatorname{sen}\phi]$$

If Newton's laws of motion are now applied, two alternative forms of the dynamic model of the quadcopter can be obtained, which are very useful for the study of the control of the vehicle.

The first, related to (76), can be expressed in matrix form as follows:

$$m\dot{\mathbf{v}} = \mathbf{R}\cdot\mathbf{f_B} = mg\hat{\mathbf{e}}_z - f_t\mathbf{R}\cdot\hat{\mathbf{e}}_3 \tag{80}$$

so that the second, related to (77), can be expressed as follows:

$$\ddot{x} = -\frac{f_t}{m}\left[\operatorname{sen}\phi\operatorname{sen}\psi + \cos\phi\cos\psi\operatorname{sen}\theta\right]$$

$$\ddot{y} = -\frac{f_t}{m}\left[\cos\phi\operatorname{sen}\psi\operatorname{sen}\theta - \cos\psi\operatorname{sen}\phi\right] \tag{81}$$

$$\ddot{z} = g - \frac{f_t}{m}\left[\cos\phi\cos\theta\right]$$

If we now take into account the consideration of Das [65] in which the vehicle experiences only small changes in its orientation angles during manoeuvres due to its low displacement velocity, the following simplification can be made $[\dot{\phi},\dot{\theta},\dot{\psi}]^T = [p,q,r]^T$, so that the dynamic model of the quadrotor in the inertial coordinate system becomes:

which allows the state vector to be redefined to the following form:

$$X = [x,y,z,\phi,\theta,\psi,\dot{x},\dot{y},\dot{z},p,q,r]^T \in \mathbf{R}^{12} \tag{83}$$

making it possible to rewrite the quadrotor equations in the state space in compact form as:

$$\dot{x} = \mathbf{f}(x) + \sum_{i=1}^{4} \mathbf{g}_i(x)u_i \qquad (84)$$

where $\mathbf{f}(x)$ and $\mathbf{g}_i$ (x) have the following form:
where for $\mathbf{g}_1$ (x) we have that:

$$g_a = -\frac{1}{m}\left[\,\text{sen}\phi\,\text{sen}\psi + \cos\phi\cos\psi\,\text{sen}\theta\right]$$

$$g_b = -\frac{1}{m}\left[\,\text{sen}\phi\,\cos\psi - \cos\phi\,\text{sen}\psi\,\text{sen}\theta\right] \qquad (87)$$

$$g_c = -\frac{1}{m}\left[\,\cos\phi\cos\theta\right]$$

4.4. Linear model in state space.

If the variable u is defined as the control vector with the following form:

$u = \left[f_t, \tau_x, \tau_y, \tau_z\right]^T \in \mathbb{R}^4.$. The linearisation procedure will develop around an equilibrium point X, which for a defined input U is the solution of the system of algebraic equations. Which can also be defined as the set of values of the state space vector which for a constant input is the solution of the algebraic system of equations, i.e.:

$$\hat{\mathbf{f}}(\overline{x}, \overline{u}) = 0 \qquad (88)$$

As the function $\mathbf{f}$ in (88) is non-linear, the problems related to the existence or not of a unique solution to the system of equations is greatly increased.

In particular, for the system of equations already established and which define the behaviour of the quadcopter, it is very difficult to find a closed solution because the trigonometric functions are related to each other in a non-elementary way. For this reason, the linearisation is developed for a simplified model where the oscillations in the posture are small, so that the simplification is based on the approximation of the sine function to its argument and the cosine function to unity.

These considerations are valid if the value in the increment of the argument is small since the vehicle is considered to be on the absolute horizontal and therefore the value of the angles is zero as stated in the Tait-Bryan angular coordinate system. The equations below describe the system resulting from the linearisation.

$$\dot{\phi} = p + r\theta + q\phi\theta$$
$$\dot{\theta} = q - r\phi$$
$$\dot{\psi} = r + q\phi$$
$$\dot{p} = \frac{I_y - I_z}{I_x} rq + \frac{\tau_x + \tau_{wx}}{I_x}$$
$$\dot{q} = \frac{I_z - I_x}{I_y} pr + \frac{\tau_y + \tau_{wy}}{I_y}$$
$$\dot{r} = \frac{I_x - I_y}{I_z} pq + \frac{\tau_z + \tau_{wz}}{I_z}$$
$$\dot{u} = rv - qw - g\theta + \frac{f_{wx}}{m}$$
$$\dot{v} = pw - ru + g\phi + \frac{f_{wy}}{m}$$
$$\dot{w} = qu - pv + g + \frac{f_{wz} - f_t}{m}$$
$$\dot{x} = w(\phi\psi + \theta) - v(\psi - \phi\theta) + u$$
$$\dot{y} = v(1 + \phi\psi\theta) - w(\phi - \psi\theta) + u\psi$$
$$\dot{z} = w - u\theta + v\phi$$

$$(89)$$

system of equations which can be written compactly with the following expression:

$$\dot{x} = f(x,u) \tag{90}$$

4.5. Linearisation in state space.

As indicated above, in order to carry out the linearisation, it is essential to define an equilibrium point, which can be the one indicated below:

$$\bar{x} = \left[0,0,0,0,0,0,0,0,0,\bar{x},\bar{y},\bar{z}\right]^{T} \in \mathbb{R}^{12} \tag{91}$$

since from the equations given in (89), it can be shown that the equilibrium condition for (91) is obtained for a constant input value defined by:

$$\bar{u} = \left[mg,0,0,0\right]^{T} \in \mathbb{R}^{4} \tag{92}$$

It should be noted that this particular value represents the force necessary to counteract the weight of the aircraft and to maintain a hovering or hovering position as a function of the disturbances due to atmospheric currents. After determining the equilibrium point x and the corresponding nominal input u, the matrices associated with the linear system can be determined and are determined by the following relationships:

$$A = \frac{\partial \mathbf{f}(\overline{x},\, \overline{u})}{\partial x} = \begin{bmatrix} 0 & 0 & 0 & 1 & 0 & 0 & 0 & 0 & 0 & 0 & 0 & 0 \\ 0 & 0 & 0 & 0 & 1 & 0 & 0 & 0 & 0 & 0 & 0 & 0 \\ 0 & 0 & 0 & 0 & 0 & 1 & 0 & 0 & 0 & 0 & 0 & 0 \\ 0 & 0 & 0 & 0 & 0 & 0 & 0 & 0 & 0 & 0 & 0 & 0 \\ 0 & 0 & 0 & 0 & 0 & 0 & 0 & 0 & 0 & 0 & 0 & 0 \\ 0 & 0 & 0 & 0 & 0 & 0 & 0 & 0 & 0 & 0 & 0 & 0 \\ 0 & -g & 0 & 0 & 0 & 0 & 0 & 0 & 0 & 0 & 0 & 0 \\ g & 0 & 0 & 0 & 0 & 0 & 0 & 0 & 0 & 0 & 0 & 0 \\ 0 & 0 & 0 & 0 & 0 & 0 & 0 & 0 & 0 & 0 & 0 & 0 \\ 0 & 0 & 0 & 0 & 0 & 0 & 1 & 0 & 0 & 0 & 0 & 0 \\ 0 & 0 & 0 & 0 & 0 & 0 & 0 & 1 & 0 & 0 & 0 & 0 \\ 0 & 0 & 0 & 0 & 0 & 0 & 0 & 0 & 1 & 0 & 0 & 0 \end{bmatrix} \qquad (93)$$

$$B = \frac{\partial \mathbf{f}(\overline{x},\, \overline{u})}{\partial u} = \begin{bmatrix} 0 & 0 & 0 & 0 \\ 0 & 0 & 0 & 0 \\ 0 & 0 & 0 & 0 \\ 0 & \dfrac{1}{I_x} & 0 & 0 \\ 0 & 0 & \dfrac{1}{I_y} & 0 \\ 0 & 0 & 0 & \dfrac{1}{I_z} \\ 0 & 0 & 0 & 0 \\ 0 & 0 & 0 & 0 \\ \dfrac{1}{m} & 0 & 0 & 0 \\ 0 & 0 & 0 & 0 \\ 0 & 0 & 0 & 0 \\ 0 & 0 & 0 & 0 \end{bmatrix} \qquad (94)$$

and if atmospheric wind disturbances are considered, defined as:

$$d = \left[f_{wx},\, f_{wy},\, f_{wz},\, \tau_{wx},\, \tau_{wy},\, \tau_{wz} \right] \in \mathrm{R}^6 \qquad (95)$$

the corresponding matrix is obtained with the form:

$$D = \frac{\partial f(\bar{x},\bar{u},d)}{\partial d} = \begin{bmatrix} 0 & 0 & 0 & 0 & 0 & 0 \\ 0 & 0 & 0 & 0 & 0 & 0 \\ 0 & 0 & 0 & 0 & 0 & 0 \\ 0 & 0 & 0 & \dfrac{1}{I_x} & 0 & 0 \\ 0 & 0 & 0 & 0 & \dfrac{1}{I_y} & 0 \\ 0 & 0 & 0 & 0 & 0 & \dfrac{1}{I_z} \\ \dfrac{1}{m} & 0 & 0 & 0 & 0 & 0 \\ 0 & \dfrac{1}{m} & 0 & 0 & 0 & 0 \\ 0 & 0 & \dfrac{1}{m} & 0 & 0 & 0 \\ 0 & 0 & 0 & 0 & 0 & 0 \\ 0 & 0 & 0 & 0 & 0 & 0 \\ 0 & 0 & 0 & 0 & 0 & 0 \end{bmatrix} \tag{96}$$

so that the linearised model is compactly defined according to (97):

$$\dot{x} = A \cdot x + B \cdot u + D \cdot d \tag{97}$$

The system of state variables is as follows:

$$\begin{aligned}
\dot{\phi} &= p \\
\dot{\theta} &= q \\
\dot{\psi} &= r \\
\dot{p} &= \frac{\tau_x + \tau_{wx}}{I_x} \\
\dot{q} &= \frac{\tau_y + \tau_{wy}}{I_y} \\
\dot{r} &= \frac{\tau_z + \tau_{wz}}{I_z} \\
\dot{u} &= -g\theta + \frac{f_{wx}}{m} \\
\dot{v} &= g\phi + \frac{f_{wy}}{m} \\
\dot{w} &= \frac{f_{wz} - f_t}{m} \\
\dot{x} &= u \\
\dot{y} &= v \\
\dot{z} &= w
\end{aligned} \tag{98}$$

4.6. Controllability and Observability of the linearised system.

Controllability and observability are the two main concepts contributed by modern control theory [66]. These concepts as such were introduced by R. Kalman in his doctoral thesis in 1960, which can be stated in general terms as follows:

A system is **controllable** at time t_0 if it is possible to transfer by using an unrestricted control vector the system from the initial state $X(t_0$)to any other state in a finite interval of time, whereby a system can be said to exhibit complete controllability if all states are controllable.

A system is **observable** at time t_0 if with the system in an initial state $X(t_0$) ; it is possible to determine this state from observations of the output $y(t)$ over a finite time interval, whereby a system can be said to exhibit complete observability if all states are observable.

The concepts of controllability and observability for a linear time-invariant dynamical system can be appropriately related to a system of linear algebraic equations of which it is widely known that the system has a solution if and only if the rank of the matrix system is complete. In this sense, the controllability and observability tests of a system can be related to the rank tests of certain matrices, which are called controllability and observability if they meet the criterion. For the purpose of studying or demonstrating the concept of observability, consider the following linear system:

$$\dot{x} = A \cdot x, \quad x(t_0) = x_0 \tag{99}$$

where A is defined according to (93) and with the following measurement equation:

$$y = C \cdot x \tag{100}$$

where the dimensions of the different variables are

$$x \in R^{12}, \quad y \in R^{12}, \quad A \in R^{12x12}, \quad C \in R^{12x12} \tag{101}$$

so that the observability matrix is given by:

$$O = \begin{bmatrix} C \\ C \cdot A \\ C \cdot A^2 \\ C \cdot A^3 \\ \vdots \\ C \cdot A^{11} \end{bmatrix} \in R^{144x12} \tag{102}$$

such that the linear time-continuous system defined by (99) and with measurements as specified in (100) is observable if and only if the rank of the matrix defined in (102) is complete.

On the other hand, the controllability matrix is defined by:

$$C = \left[B, \ A \cdot B, \ A^2 \cdot B, \ A^3 \cdot B, \cdots, \ A^{11} \cdot B \right] \in \mathbb{R}^{12 \times 48} \qquad (103)$$

where B is defined in (94), and the linear time-continuous system is defined in (99) with measurement equation defined in (100). So this system is controllable if and only if the controllability matrix has full rank.

From a practical point of view and because of the size of the matrices involved, the verification of the controllability and observability of the quadcopter model will be performed using MATLAB.

4.7. Position control by means of a Linear Quadratic Regulator (LQR).

The objective of the optimal control techniques according to Murray [67] is to determine the necessary control signals in such a way that in order for the system to be controlled, physical constraints are met to minimise the cost function and/or maximise the performance function. Specifically, the solution of an optimisation problem is to try to take the system state $*(t)$ along the desired trajectory $*_d$ while minimising the cost of the transfer. In addition, it should minimise the use of control inputs, thus reducing the use of actuators.

In addition to the above, the secondary needs in the optimisation of control are:

< A model capable of describing in the best possible way the behaviour of the dynamic system under control.

< A cost index J , which takes into account the specifications of the system and the needs of the designer.

< Definition of possible boundary conditions and physical restrictions that limit the system.

Consider a dynamic system where the state variables belong to the set *, and the control inputs belong to the set u, such that:

$$\dot{x}(t) = f\left[x(t), u(t), t \right] \qquad (104)$$

where x is a vector of length $n-1$, and u is a vector of length $r-1$, so that it is possible to define the cost function, J, as:

$$J = e\left[x(tf) \right] + \int_{t_0}^{t_f} w\left[x(t), u(t), t \right] dt \qquad (105)$$

where the terminal cost, e, is a non-negative function as is the weight function, w, such that $e(0) = 0, y \ w(0, 0, t) = 0,$, where the boundary conditions are:

$x(t_0) = x_0, \ x(t_f) \, y \, t_f$ and t_f without constraints. It is possible to define the solution

47

of the optimisation problem as:

$$u(t), \forall t \in \left[t_0, t_f\right] \qquad (106)$$

which is a function that seeks to minimise J. Considering that the limit of the time interval is an infinite approximation, the system and the cost index manifest themselves as follows:

$$\begin{cases} \dot{x} = A \cdot x + B \cdot u \\ y = C \cdot x \end{cases} \qquad (107)$$

$$J = \int_{t_0}^{\infty} \left\{ u(t)^T \cdot R \cdot u(t) + \left[x(t) - x_d(t)\right]^T \cdot Q \cdot \left[x(t) - x_d(t)\right] \right\} dt \qquad (108)$$

being $R \, y \, Q$ matrices that:

$\mathbf{R}$ is the cost of the actuators $(R = R^T$ which is a positive definite matrix; $R \in R^{m \times m})$.

$\mathbf{Q}$ is the cost of the states $(Q = Q^T$ which is a positive semi-defined matrix; $Q \in R^{n \times n})$.

And as shown by Murray [70] the control input *u(t)* which minimises the functional is a linear feedback of states of the form:

$$u(t) = -K \cdot \left[x(t) - x_d(t)\right] \qquad (109)$$

where:

$$K = R^{-1} \cdot B^T \cdot S \qquad (110)$$

S being the matrix which is the solution of the algebraic Ricatti equation, given below:

$$S \cdot A + A^T \cdot S - S \cdot B R^{-1} \cdot B^T \cdot S + C^T \cdot Q \cdot C = 0 \qquad (111)$$

where S is a positive definite matrix, Figure 13 being the schematic of the implemented system.

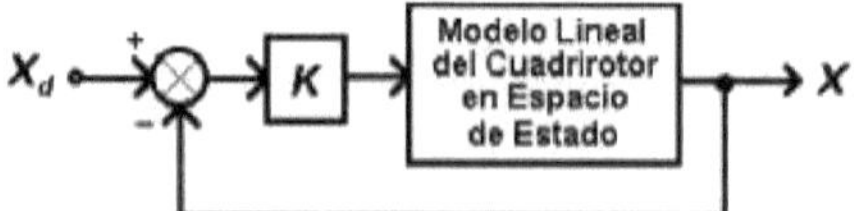

Figura 13 Esquema de control LQR
Figure 13 LQR control schematic

It should be noted that although the control scheme may seem very simple, it is a matrix system with the vector x_d defining the coordinates of the desired spatial position and the vector x defining the spatial position obtained as a function of time, i.e. the path travelled to get from one position to the other.

It should be noted that the algebraic equation can be solved by means of the Ricatti method using the MATLAB function LQR, the syntax of which is defined below:

$$K = LQR(A, B, Q, R) \tag{112}$$

According to the work of Murray [70], on the LQR controller problem for an infinite horizon and the factorisation of the matrix $Q = E^T \cdot E$, if the pair formed by the matrices A and B is controllable and the pair formed by the matrices A and E is observable, it can be stated that:

< There must exist a matrix S which is a positive definite solution of the algebraic Ricatti equation.

< The closed-loop system $\dot{x} = (A - BK) \cdot x$ is asymptotically stable, with $K = R^{-1} \cdot B^T \cdot S$.

A simple numerical test that allows to choose the values of the matrices Q and R taking into account the matrices A (93) and B (94) is achieved by considering the linear system defined in (97) without considering the part corresponding to the perturbations by the atmospheric wind gusts.

Obviously, taking into account the complexity of the equations due to their matrix structure, the simulation system will be developed using MATLAB software and all its associated functions by means of a command line script as it allows greater control in terms of sequence and structure of operation in relation to the use of SIMULINK.

4.8. Structure of the simulation programme.

As such, the simulation program developed for the modelling of the quadcopter and its performance with the LQR controller in the tracking of any trajectory has an extension of more than 800 lines of code in MATLAB instructions, for which reason it is not annexed to the present document, but nevertheless, the general structure or order of sequence of activities for the achievement of the proposed goal is indicated below.

First of all and as a common factor of all computer programs, the workspace and memory were "cleared" of all numerical and graphical variables. Next, the graphic resolution parameters of the monitor were defined, since obviously, being the main work a simulation, all the results must be visualised. The coordinates of the different spatial points of interest in the trajectory to be traversed were then defined in a matrix format of n rows x 3 columns. The advantage of this matrix format is that several trajectories can be defined for the purpose of observing behavioural results. To finalise this initial configuration segment of the program and taking into account the large number of graphical results, a menu was created to define the results to be displayed in each run of the program.

After the initial configuration, the physical dimensional values of the quadcopter are loaded and the anahetic model of linearised equations that describe the device is developed and in which the dimensional values will be substituted during the simulation. At this point it is necessary to verify the matrices in the state space that model the quadcopter. The stability of the continuous mathematical model of the system is verified by observing its response to a unitary step input. The controllability and observability matrices of the continuous system are also verified by calculating the range of these matrices.

Once the mathematical functionality of the various matrices describing the aircraft model has been verified, and taking into account that MATLAB as a simulation system works in time increments, it is decided to discretise the linear continuous system in order to use the same sampling period for all the calculations to be performed, and in all the functions that need to be used.

With the system of discretised equations, the verification of the matrices in the state space that model the four-engine is carried out again. The stability of the mathematical model of the system is verified by observing its response to a unitary step input. The controllability and observability matrices are also verified by calculating the range of these matrices, which must be equal in numerical result to those of the continuous system for correspondence purposes.

Finally, the results of the three-dimensional navigation, the behaviour of the state variables, and the control actions necessary to follow the trajectory defined at the beginning by means of its points of interest are visualised in order to check the performance of the LQR controller.

5.1. Results of the simulation.

Below is a series of images resulting from the graphing of the numerical values of both the continuous and discrete open-loop mathematical model describing the vehicle, and the closed-loop system using an LQR controller that allows the simulation of the quadcopter flight in the simplest trajectory when from a point of origin with coordinates x=0, y=0, and z=0 it goes to a single point of arrival with coordinates x = 5.0, y = 5.1, z = 5.2.

Figure 14 shows the map of poles and zeros in the S-plane of the linearised continuous mathematical model.

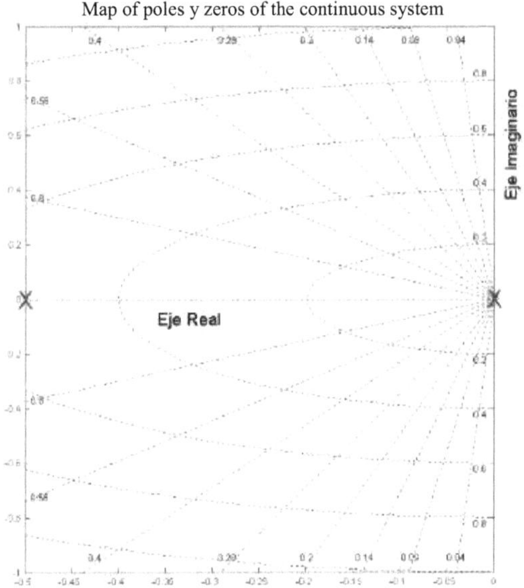

Figure 14 Open-loop pole-zero map of the linearised system.

In Figure 15, the response of the open-loop system to a unitary step input can be seen. The reason for choosing this type of input is based on the fact that the navigation trajectory is defined by separate points of interest, so that the quadrotor's journey from one point of interest to another can be considered as a three-dimensional non-unitary step change.

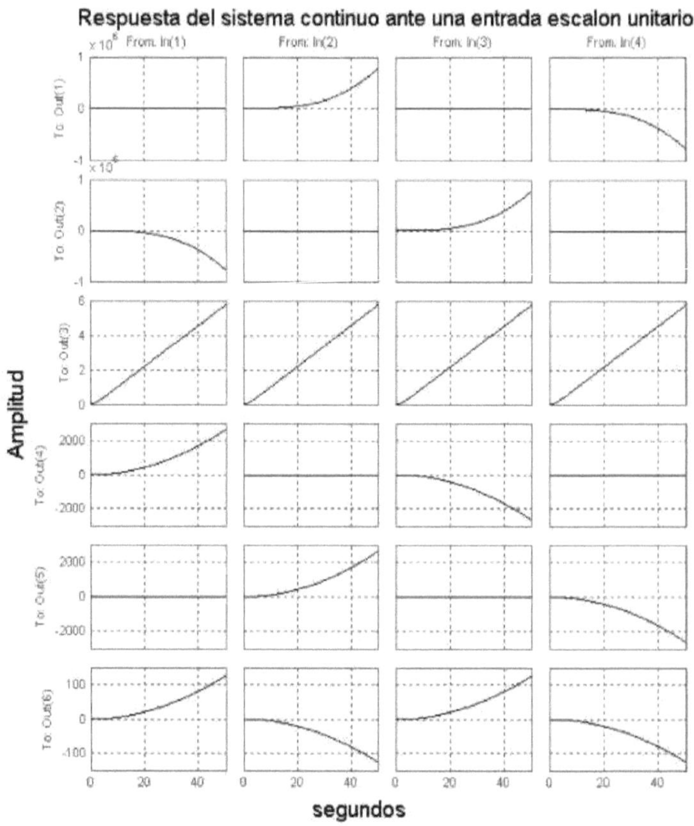

Figura 15 Respuesta al escalón unitario en lazo abierto.

Figure 16 shows the response provided by MATLAB when queried about the value of the poles of the system in continuous time, open loop, and about the ranges of the controllability and observability matrices in order to verify whether the linearisation considerations allow the vehicle to be controlled and its state variables to be observed directly.

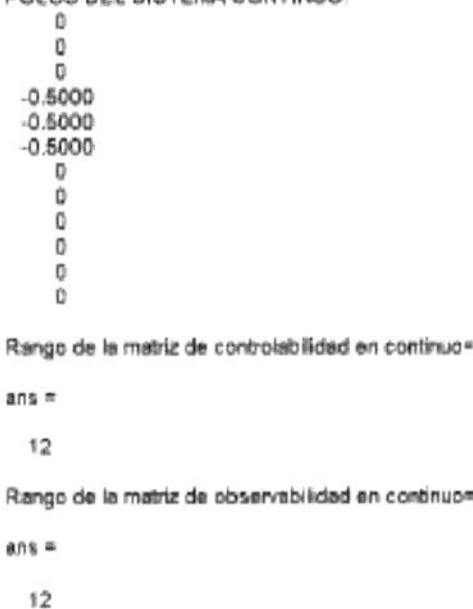

Figure 16 MATLAB response on the location of the poles and ranges of the matrices.

Figure 17 shows the map of poles and zeros in the Z-plane of the mathematical model of the discretised system with a sampling period of 0.1 seconds.

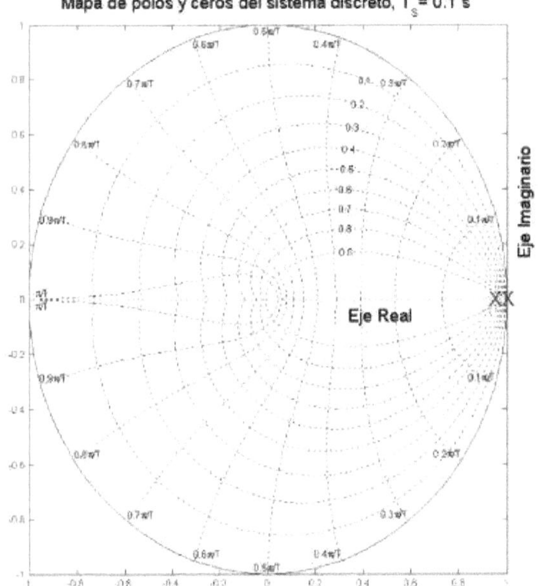

Figure 17 Map of poles and zeros of the discrete system.

Figure 18 shows the response of the discretised system with a sampling period of 0.1 seconds and in open loop to a unit step input.

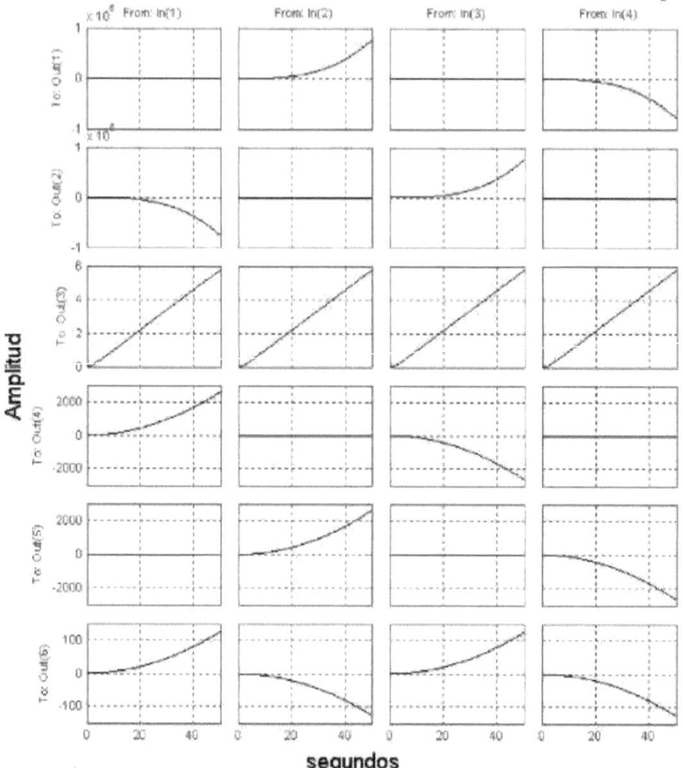

Figura 18 Respuesta del sistema discreto a una entrada escalón unitario.

Figure 19 shows the response provided by MATLAB when queried about the value of the poles of the system in discrete time, open loop, and about the ranges of the controllability and observability matrices in order to verify whether the linearisation considerations allow the vehicle to be controlled and its state variables to be observed directly.

```
POLOS DEL SISTEMA DISCRETO:
  1.0000
  1.0000
  1.0000
  0.9512
  0.9512
  0.9512
  1.0000
  1.0000
  1.0000
  1.0000
  1.0000
  1.0000

Rango de la matriz de controlabilidad en discreto=

ans =

  12

Rango de la matriz de observabilidad en discreto=

ans =

  12
```

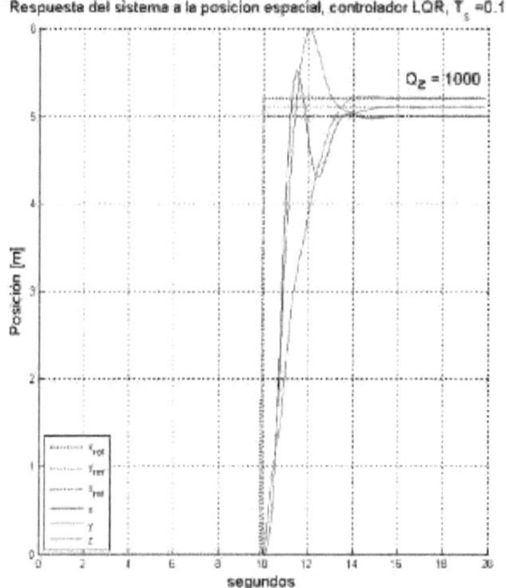

Figure 19 MATLAB response on the location of the poles and ranges of the matrices.

Figure 20 shows the response of the quadcopter to the positional change coordinates resulting in the navigation of the vehicle with a penalty value for the Z-coordinate axis state variable of 1000 in the cost matrix Q.

Figure 20 Response to trajectory tracking.

Figure 21 shows the displacement velocities with respect to the inertial coordinate system that the quadrotor undergoes in order to achieve the spatial position change, with a penalty value of 1000 for the state variable of the Z coordinate axis in the cost matrix Q.

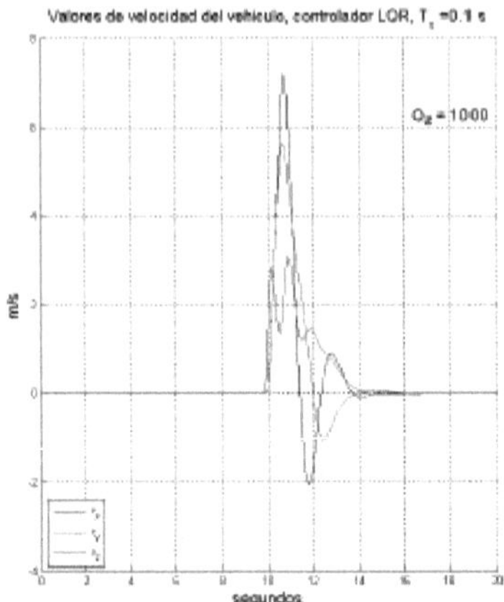

Figure 21 Velocities of the vehicle in the inertial coordinate system

Figure 22 shows the value of the angles defining the quadcopter's posture in the vehicle or mobile coordinate system, with a penalty value for the Z coordinate axis state variable of 1000 in the cost matrix Q.

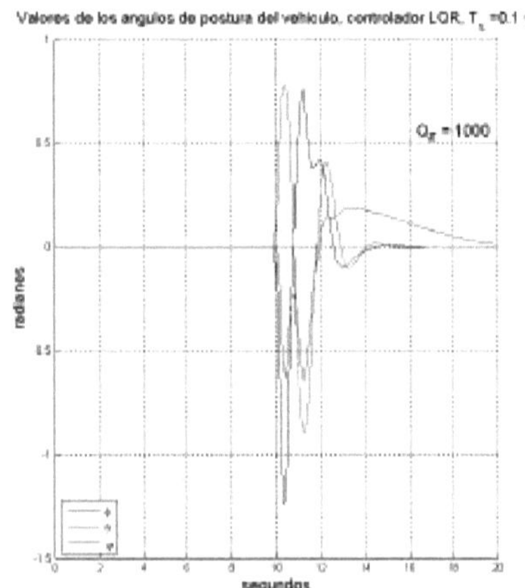

Figure 22 Quadrotor posture values.

Figure 23 shows the values of the rotations experienced by the air vehicle on its y-axes with respect to the inertial coordinate system, with a penalty value of 1000 for the state variable of the Z-coordinate axis in the cost matrix Q.

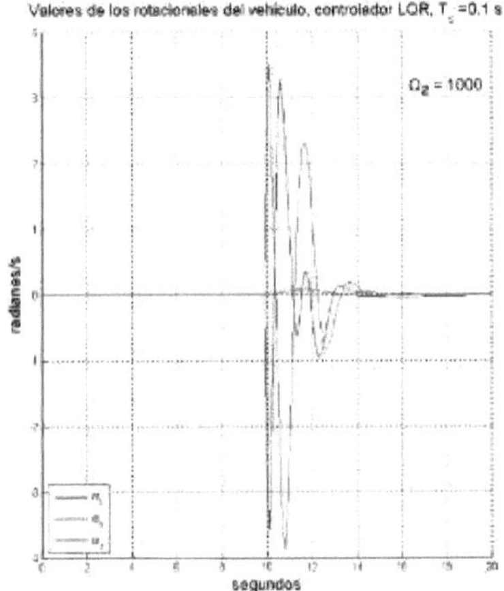

Figure 23 Aircraft rotational values.

Figure 24 shows the spatial navigation performed by the aerial vehicle, with a penalty value for the Z-coordinate axis state variable of 1000 in the cost matrix Q.

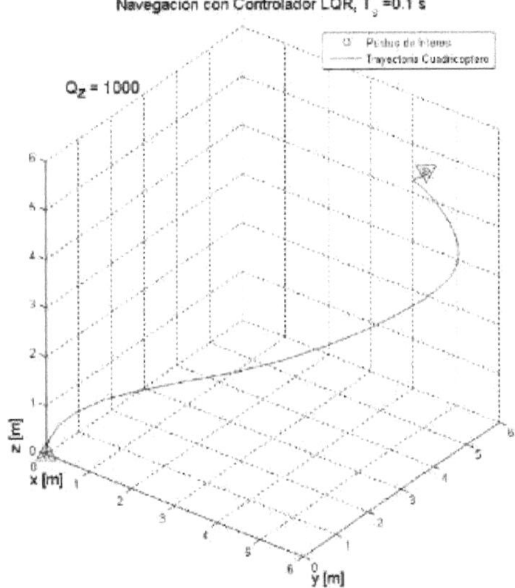

Figure 24 Trajectory realised.

Figure 25 shows the rotational velocity that the various propellers must have to perform the navigation described in Figure 24 with a penalty value of 1000 for the Z coordinate axis state variable in the cost matrix Q.

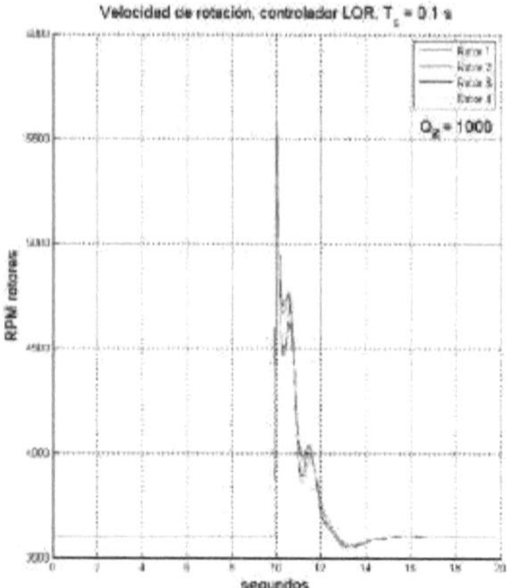

Figure 25 Control action values on the propellers.

The following is a series of figures for the same complex trajectory to check the real performance of the simulation system by increasing the penalty value only for the cost matrix Q, which is related to the system states, so that undesired states can be penalised in order to reach the steady state as soon as possible. These penalty values are a way of modifying the values of the gain matrix K in order to tune the controller.

As for the cost matrix R, which is related to the control actions, it was decided to assign it a value of unity, the identity matrix, since increasing the weights (keeping the weights of Q constant) increases the penalisation by more energetic control actions or with higher energy consumption, so that the control efforts are smaller as well as the values of the gain matrix K, and for this reason the system reaches the steady state more slowly.

From the above, it is established that for the purposes of the simulation system, the control assumption to be met is that the system stabilises as quickly as possible, fast response, regardless of the energy requirements of the control actions.

As already indicated, the cost matrix Q corresponds to the state variables, of which there are twelve, and the cost matrix R corresponds to the control action by means of the propellers, of which there are four. As can be inferred, the number of combinations grows exponentially, which is beyond the scope of this work. In this sense, as the object of an aerial vehicle is to lift off by overcoming the effect of weight through the action of the acceleration of gravity, only the penalisation for the state variable corresponding to the z-axis in the cost matrix Q is increased.

The weight values that Qz will take in order to increase the penalty are: 1, 10,

100, 1000, 10000, and 100000, which means that in each case the airborne system will be less tolerant to deviations in the steady-state vehicle elevation. The values of the navigation coordinates for the complex trajectory are given in Table 6.1.

Table 6.1. Navigation coordinate values, in metres.

X	Y	Z	Observation
0	0	0	Start-up
0	0	1.8	
0	2.3	1.8	
2.2	2.1	1.8	
2.7	-1.8	1.3	
-2.3	-2.7	1.8	
-2.4	1.3	0.8	
-1.75	1.1	0	Arrival

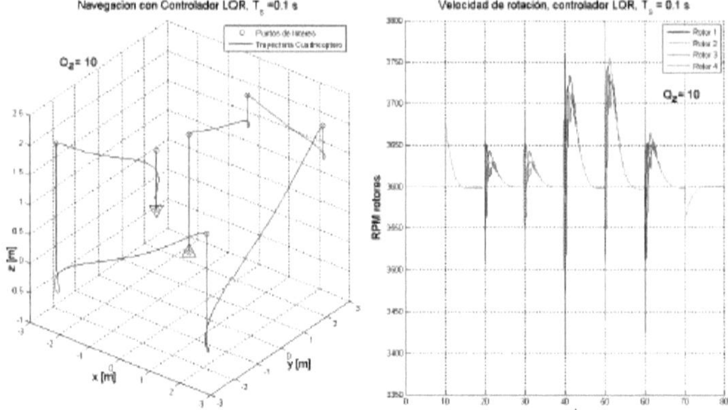

Figura 26 Navegación con Qz=10.

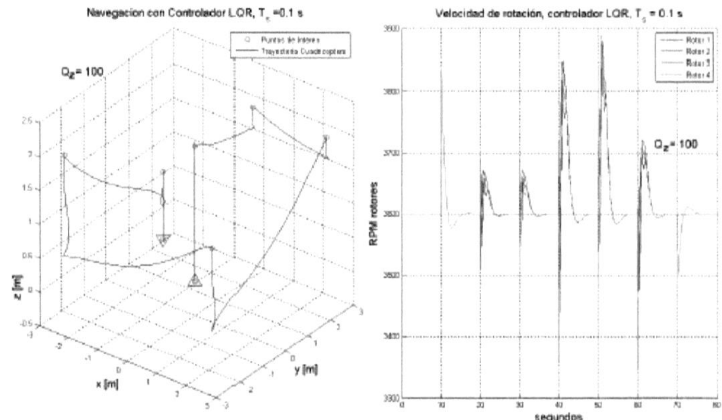

Figura 27 Navegación con Qz=100.
Figure 26 Navigation with Qz=10.
Figure 27 Navigation with Qz=100.

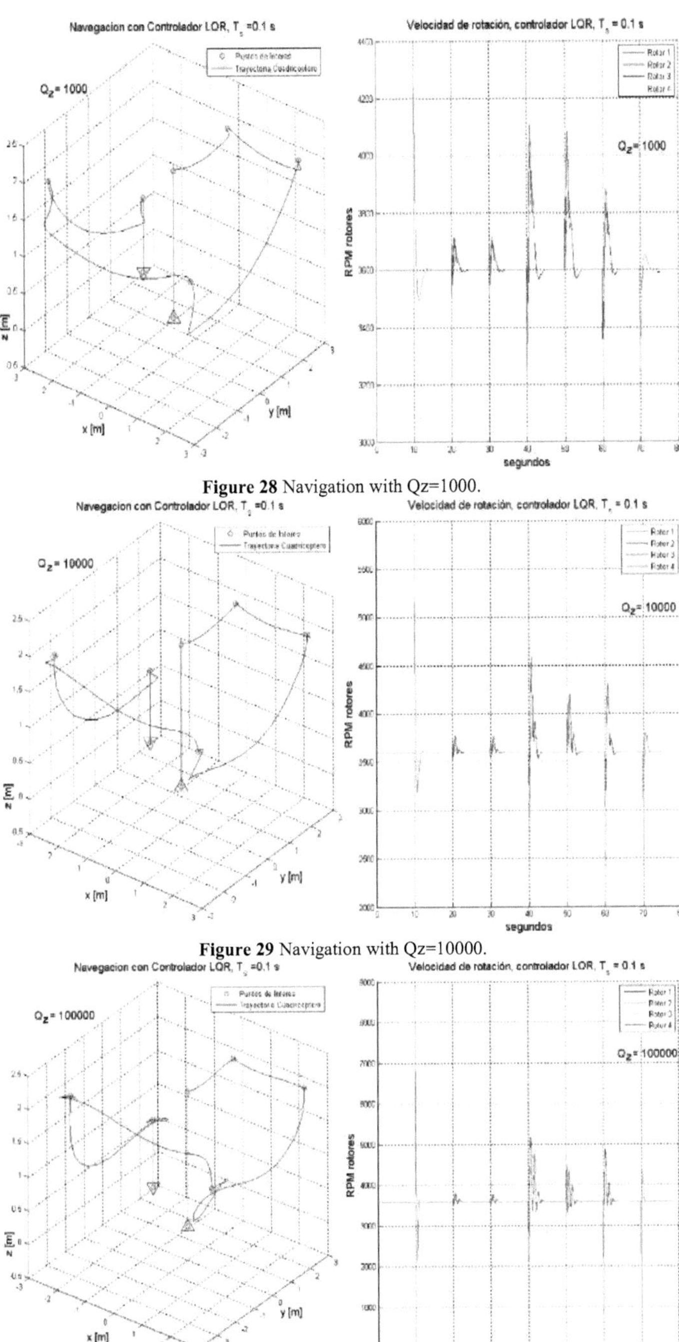

Figure 28 Navigation with Qz=1000.

Figure 29 Navigation with Qz=10000.

Figure 30 Navigation with Qz=100000.

5.2. Analysis of the results of the simulation.

Figure 14 shows that the continuous, linearised, open-loop system has only two poles, a second-order system, and one of the poles is at the origin, which establishes that it is a marginally stable system.

Figure 15, in contrast to Figure 14, shows that the open-loop system is neither second-order nor marginally stable, as several of the system's responses to a unit step input are unbounded.

Figure 16 shows the response of MATLAB when asked about the value of the poles of the open-loop system, as well as the range of the controllability and observability matrices. As for the value of the poles of the system, it can be seen that there are nine poles on the origin, which establishes without a doubt that the system is unstable by its own nature, since the multiplicity of poles on the imaginary axis is greater than one. As for the ranges of the matrices, it can be seen that the response is 12 for each one of them, which indicates that the number of linearly independent equations is equal to the number of state variables, which is why it can be established that the system is controllable and observable in the loop.
closed.

For the case of Figure 17, the continuous system has been discretised with a sampling period of 0.1 second for proper simulation in MATLAB, showing that this open-loop model has poles on the unit circle indicating that the discrete open-loop system is unstable by its own nature.

In Figure 18, the instability of the discrete open-loop system is proven by the fact that several of the responses to a unit step input are unbounded.

Figure 19 shows the response of MATLAB when asked about the value of the poles of the discrete open-loop system, as well as the range of the controllability and observability matrices. As for the value of the poles of the system, it can be seen that there are nine poles above the unit circle, which establishes without a doubt that the system is unstable by its own nature. As for the ranges of the matrices, it can be seen that the response is 12 for each one of them, which indicates that the discrete system is observable and controllable in a closed loop.

Figure 20 shows the response of the system to the tracking of a simple trajectory, in which the closed-loop system with LQR controller has a certain level of overshoot in the transient regime but zero position error in the steady state. On the other hand, it can be seen that the system has a stabilisation time of about four seconds, which indicates that the time constant of the closed-loop system is about one second. The important point to note here is that the system is controllable in the closed loop even though it is unstable in the open loop.

Figure 21 shows the velocities developed by the air vehicle to move in space from point A (0, 0, 0) to point B (5.0, 5.1, 5.2) with respect to the inertial coordinate system.

Figure 22 shows the values of the angles describing the vehicle's position in the moving coordinate system.

Figure 23 shows the angular velocity values experienced by the vehicle on its axes with respect to the inertial coordinate system.

In Figure 24, the trajectory followed by the quadrotor to go from point A to point B is shown in three dimensions. While it is clear that the shortest distance between two points is a straight line, with the values defined in the matrices Q and R, cost of operation, the optimal trajectory for these considerations is a spiral arc since only deviations in the Z-axis are penalised.

Figure 25 shows the angular velocity values, in revolutions per minute, that the final control elements, propellers or rotors, had to undergo in order to navigate the trajectory from point A to point B with a weight value of 1000 as a penalty in Qz.

Figure 26 to Figure 31 show the results for the navigation of a complex trajectory where only the penalty on the Z-axis for the cost matrix Q is modified. For the specific case of Figure 26, the penalty is so low (Qz=1) that the positioning errors are very significant and the control efforts (rotational speed of the propellers) are so low that the vehicle has a stabilisation time of more than 10 seconds, which means that it does not even reach the points of interest of the trajectory.

As for Figure 27, the penalty weight is increased to a value of 10 and although the positioning errors are still significant, the increase in control effort reduces the stabilisation time so that at least the trajectory points of interest are reached.

In relation to Figure 28, the value of the weight in the Qz penalty is increased to 100, which reduces the positioning error and improves the time response of the system. Obviously, increasing the control effort.

For the case of Figure 29, now the weight value in the Qz penalty is 1000 resulting in a significant reduction in the positioning error with moderate control effort and in which the time constant of the system is reduced to approximately one second, so that the vehicle takes advantage of the inertia due to the effect of gravity acceleration to reach the next point of interest in the defined trajectory.

As for Figure 30, the value of the weight in Qz is 10000, which means that the penalty for position errors is large, producing an even greater reduction in the vehicle's position errors but with a significant increase in the control effort, which does not necessarily translate into a reduction in the system's time constant but does result in greater energy consumption. This condition is ignored because of the established control premise, according to which the important thing is that the system stabilises or reaches the point of interest in the shortest possible time.

Referring to Figure 31, the penalty in Qz is the largest of all when using a weight value of 100000. However, the reduction in positioning error is not significant compared to the cases where Qz=10000 or Qz=1000. On the other hand, a Qz=100000 implies such a large control effort that the rotational speed of the propellers could well be prohibitive either because of the mechanical resistance to traction, the possibility of the motor not reaching it, or the rapid depletion of the battery that powers the quadcopter.

From the results obtained from Figures 29, 30, and 31, it is inferred that there is an inverse relationship between the Q and R matrices, since increasing the penalty in Qz for the deviation in the position state variable to be minimal, in turn requires an increase in the control effort or control actions more energetic in magnitude that could not necessarily be carried out by the final control elements. The above conjecture allows us to infer that in optimal control the main line of investigation must be the establishment of at least one methodology that allows us to define the values of the Q and R matrices most appropriate to the physical system to be manipulated according to the desired control assumptions.

6.1.Conclusions.

From the Euler equations it was possible to develop the mathematical model that describes the behaviour and performance of the quadcopter that is expected to simulate and control its flight trajectory.

By linearising the mathematical equations in a system of state variables it was possible to determine in both continuous and discrete time the ranges of the controllability and observability matrices, thus determining that in both time domains it was possible to observe the state variables of the aeronautical system and control its flight trajectory.

Using the MATLAB numerical analysis program, it was possible to develop the simulation system, in which, for any trajectory defined by n points of interest, the performance of the airspeed in the tracking of this trajectory was controlled.

The fact that the gain matrix K is not dynamic since it depends on the weights in the Q and R matrices, makes it necessary to test with different penalty values according to the control premise sought, thus tuning the controller to increase the quality of its design.

The evaluation of the performance of the LQR controller on a complex trajectory indicates how decisive the appropriate choice of the values of the Q and R matrices is for the performance of the air vehicle, as they determine the suitability or otherwise of the flight behaviour for the human operator.

BIBLIOGRAPHICAL REFERENCES

[1] Description of the helicopter created by Etienne Oemichen. Available in: http://www.aviastar.org/helicopters eng/oemichen.php. Revised: 06 June 2016.

[2] Description of the helicopter created by George Bothezat. Available at: http://www.aviastar.org/helicopters eng/bothezat.php. Revised: 06 June 2016.

[3] S. Raza and Wail Gueaieb, Intelligent Flight Control of an Autonomous Quadrotor, Motion Control, InTech, University of Ottawa, Canada, 2010. Available in: http://cdn.intechopen.com/pdfs/6587/intech-intelligent flight control of an autonomous quadroto r.pdf. Revised: 06 June 2016.

[4] Eduard Santamaria , Florian Segor, Igor Tchouchenkov, Rapid Aerial Mapping with Multiple Heterogeneous Unmanned Vehicles. Available at: http://www.iscramlive.org/ISCRAM2013/files/211 .pdf. Revised: 06 June 2016.

[5] Sea rescue drone. Available at: http://www.wired.co.uk/article/iranian-rescue-robot. Revised: 06 June 2016.

[6] Alex Kushleyev, Daniel Mellinger, Vijay Kumar, GRASP Lab, University of Pennsylvania. A Swarm of Nano Quadrotors. Available at: http://www.youtube.com/watch?v=YQIMGV5vtd4. Revised: June 06, 2016.

[7] Hongning Hou, Jian Zhuang, Hu Xia, Guanwei Wang, and Dehong Yu. A simple
controller of minisize quad-rotor vehicle. In Mechatronics and Automation (ICMA), 2010 International Conference on, pages 1701-1706, 2010.

[8] Jinhyun Kim, Min-Sung Kang, and Sangdeok Park. Accurate modeling and robust hovering control for a quadrotor vtol aircraft. Journal of Intelligent and Robotic Systems, 57(1-4):9-26, 2010. ISSN 0921-0296. doi: 10.1007/s10846-009-9369-z. URL http://dx.doi.org/10.1007/s10846-009-9369-z.

[9] Randal W. Beard. Quadrotor Dynamics and Control. Lecture notes. Brigham Young University. Available at: http://scholarsarchive.byu.edu/cgi/viewcontent.cgi?article=2324&context=fac pub. Revised: June 06, 2016.

[10] Teppo Luukkonnen, Modelling and control of quadrocopter, Independent research project in applied mathematics, Aalto University, Espoo, Finland 2011. Available at: http://sal.aalto.fi/publications/pdf-files/eluu11jpublic.pdf. Revised: 06 June 2016.

[11] Samir Bouabdallah, Roland Siegwart. Full Control of a Quadrotor. Available at: https://e-collection.library.ethz.ch/eserv.php?pid=eth:7848&dsID=eth-7848-0 1 .pdf. Revised: 06 June 2016.

[12] G., Hoffmann. Huang, S. Waslander, and C. Tomlin, "Quadrotor helicopter flight dynamics and control: Theory and experiment," in Proc. of the AIAA Guidance, Navigation, and Control Conference, 2007, pp. 1-20.

[13] R. Goela, S. Shahb, N. Guptac, and N. Ananthkrishnanc, "Modeling, simulation and flight testing of an autonomous quadrotor," Proceedings of ICEAE, 2009.

[14] B. Erginer and E. Altug, "Modeling and pd control of a quadrotor vtol vehicle," in Intelligent Vehicles Symposium, 2007 IEEE. IEEE, 2007, pp. 894-899.

[15] R.V. Jategaonkar. Flight vehicle system identification: a time domain methodology. American Institute of Aeronautics and Astronautics, 2006.

[16] I.D. Cowling, O.A. Yakimenko, J.F. Whidborne, and A.K. Cooke. A prototype of an autonomous controller for a quadrotor uav. In European Control Conference, pages 1-8, 2007.

[17] S. Bouabdallah, A. Noth, and R. Siegwart. PID vs LQ control techniques applied to an indoor micro quadrotor. In International Conference on Intelligent Robots and Systems 2004, volume 3, pages 2451 - 2456 vol.3, 2004.

[18] P. Pounds, R. Mahony, and P. Corke. Modelling and control of a quadrotor robot. In Australasian conference on robotics and automation 2006, Auckland, NZ, 2006.

[19] I. Cowling, J. Whidborne, and A. Cooke, "Optimal trajectory planning and lqr control for a quadrotor uav," in Proceedings of the International Conference Control-2006, Glasgow, Scotland, vol. 30, 2006.

[20] L. Minh and C. Ha, "Modeling and control of quadrotor mav using vision-based measurement," in Strategic Technology (IFOST), 2010 International Forum on. IEEE, 2010, pp. 70-75.

[21] Y. Al-Younes, M. Al-Jarrah, and A. Jhemi, "Linear vs. nonlinear control techniques for a quadrotor vehicle," in Mechatronics and its Applications (ISMA), 2010 7th International Symposium on. IEEE, 2010, pp. 1-10.

[22] S. Bouabdallah, "Design and control of quadrotors with application to autonomous flying" Ph.D. dissertation, Federal Polytechnic School of Lausanne, 2007.

[23] D. Mellinger, Q. Lindsey, M. Shomin, and V. Kumar, "Design, modeling, estimation and control for aerial grasping and manipulation," in Intelligent Robots and Systems (IROS), 2011 IEEE/RSJ International Conference on. IEEE, 2011, pp. 2668-2673.

[24] E. Altug, J. Ostrowski, and R. Mahony, "Control of a quadrotor helicopter using visual feedback," in Robotics and Automation, 2002. Proceedings. ICRA'02. IEEE International Conference on, vol. 1. IEEE, 2002, pp. 72-77.

[25]   T. Madani and A. Benallegue, "Control of a quadrotor mini-helicopter via full state backstepping technique," in Decision and Control, 2006 45th IEEE Conference on. IEEE, 2006, pp. 1515-1520.

[26]   T. Madani and A. Benallegue. Backstepping Control for a Quadrotor Helicopter. In Intelligent Robots and Systems, 2006 IEEE/RSJ International Conference on, pages 3255 -3260, 2006.

[27]   S. Bouabdallah and R. Siegwart. Backstepping and sliding-mode techniques applied to an indoor micro quadrotor. In International Conference on Robotics and Automation 2005, pages 2247 - 2252, april 2005.

[28]   R. Xu and U. Ozguner. Sliding mode control of a quadrotor helicopter. In Decision and Control, 2006 45th IEEE Conference on, pages 4957 -4962, dec. 2006.

[29]   D. Lee, H. Jin Kim, and S. Sastry. Feedback linearization vs. adaptive sliding mode control for a quadrotor helicopter. International Journal of Control, Automation and Systems, 7(3):419-428, 2009.

[30]   A. Mokhtari, N. M'Sirdi, K. Meghriche, and A. Belaidi, "Feedback linearization and linear observer for a quadrotor unmanned aerial vehicle," Advanced Robotics, vol. 20, no. 1, pp. 71-91, 2006.

[31]   A. Das, K. Subbarao, and F. Lewis, "Dynamic inversion with zero-dynamics stabilisation for quadrotor control," Control Theory & Applications, IET, vol. 3, no. 3, pp. 303-314, 2009.

[32]   M. Hehn and R. DAndrea, "Quadrocopter trajectory generation and control," in Proceedings of the IFAC world congress, 2011.

[33]   B. Whitehead and S. Bieniawski. Model Reference Adaptive Control of a Quadrotor UAV. In Guidance Navigation and Control Conference 2010, Toronto, Ontario, Canada, 2010. AIAA.

[34]   M.Huang, B.Xian, C.Diao, K.Yang, and Y.Feng. Adaptive tracking control of underactuated quadrotor unmanned aerial vehicles via backstepping. In American Control Conference (ACC), 2010, pages 2076 -2081, 30 2010-july 2 2010.

[35]   W. Zeng, B. Xian, C. Diao, Q. Yin, H. Li, and Y. Yang. Nonlinear adaptive regulation control of a quadrotor unmanned aerial vehicle. In Control Applications (CCA), 2011 IEEE International Conference on, pages 133 -138, 2011.

[36]   J.J. Craig, P. Hsu, and S.S. Sastry. Adaptive control of mechanical manipulators. The International Journal of Robotics Research, 6(2):16, 1987.

[37]   S. Bouabdallah, P. Murrieri, and R. Siegwart. Design and control of an indoor micro quadrotor. In Robotics and Automation, 2004. Proceedings. ICRA'04. 2004 IEEE International Conference on, volume 5.

[38]   Becker, M. ; Bouabdallah, S. ; Siegwart, R. (2006). Desenvolvimento de um Controlador de Desvio de Obstaculos para um Mini-Helicoptero Quadrirotor Autonomo 1^a Fase: Simulagao. In proc. of CBA 2006 - Congresso Brasileiro de Automatica, Salvador - BA, Brazil, Vol. 1. pp. 1201-1206.

[39]   P. Pounds, R. Mahony, J. Gresham, P. Corke, and J. Roberts. Towards Dynamically-Favourable Quad-Rotor Aerial Robots. 2004.

[40]   P. Pounds, R. Mahony, and P. Corke. Modelling and Control of a Quad-Rotor Robot. 2006.

[41]   A. Mokhtari, A. Benallegue, and B. Daachi. Robust feedback linearization and gh8 controller for a quad rotor unmanned aerial vehicle. Journal of Electrical Engineering, 57(1):20-27, 2006.

[42]   G.P. Tournier, M. Valenti, J.P. How, and E. Feron. Estimation and control of a quadrotor vehicle using monocular vision and moire patterns. 2006.

[43]   E. Courses and T. Surveys. Robust low altitude behavior control of a quadrotor rotorcraft through sliding modes. pages 1-6, 2007.

[44]   Katie Miller. Path tracking control for quadrotor helicopters. July 2008.

[45]   Gjioni E. AbouSleiman R., Korff D. and Yang H. The oakland university unmanned aerial quadrotor system. 2008.

[46]   J. Gordon Leishman. A History of Helicopter Flight. Available at: http://itlims.meil.pw.edu.pl/zsis/pomoce/WTLK/ENG/Sup/A History of Heli copter Flinht.pdk Revised: 25 April 2016.

[47]   B e l l B o eing Quad Ti l tRot or . Di sp onib l e at : https://en.wikipedia.org/wiki/BellBoeingQuadTiltRotor. Revised: 25 April 2016.

[48]   Bell QTR Quad Tiltrotor project. Available at: http://www.aviastar.org/helicopters eng/bell qtr.php. Revised: 25 April 2016.

[49]   D iversity in design. Available at: http://www.boemg.com/news/frontiers/archive/2006/december/i ids03.pdf. Revised: 25 April 2016.

[50]   Moller Skycar M400, January 2009. Available at: http://en.wikipedia.org/wiki/Moller Skycar M400. Revised: 25 April 2016.

[51]   Mo l l e r M2 0 0 G V o l ant o r . Di sp oni b l e at : https://en.wikipedia.org/wiki/Moller M200G Volantor. Revised: 25 April 2016.

[52]   Beard, Randal, "Quadrotor Dynamics and Control Rev 0.1" (2008). All Faculty Publications. Paper 1325. Available at: http://scholarsarchive.byu.edu/facpub/1325. Revised: June 28, 2016.

[53]   B. Siciliano, L. Sciavicco, L. Villani, G. Oriolo. Robotics. McGraw-Hill.

[54]   D. Lee, T. Burg, D. Dawson, D. Shu, B. Xian, and E. Tatlicioglu, "Robust

tracking control of an underactuated quadrotor aerial-robot based on a parametric uncertain model", in Systems, Man and Cybernetics, 2009. SMC 2009. IEEE International Conference on, 2009, pp. 3187-3192.

[55]   Katsuhiko Ogata. Ingenieria de control moderna. Pearson Education S.A, Madrid 2010. ISBN: 978-84-8322-660-5

[56]   Julio H. Braslavsky. Introduction to optimal control. Lecturer notes 2001. Available in: http://www.eng.newcastle.edu.au/~jhb519/teaching/caut2/classes/Cap9.pdf. Revised: 05 July 2016.

[57]   John S. Bay. Fundamentals of Linear State Space Systems. WCB/McGraw-Hill, 1999.

[58]   Green, Limebeer: Linear Robust Control.

[59]   Van Willigenburg L.G., De Koning W.L. (1999). "Optimal reduced-order compensators for time-varying discrete-time systems with deterministic and white parameters". Automatica 35: 129-138.

[60]   G. Done and D. Balmford, Bramwell's Helicopter Dynamics. Oxford Butterworth-Heinemann, 2001.

[61]   P. Mullhaupt, Analysis and Control of Underactuated Mechanical Nonminimum-phase Systems. PhD thesis, EPFL, 1999.

[62]   H. Goldstein et al., Classical Mechanics. Addison Wesley, 2002.

[63]   B. Etkin and L. Reid, Dynamics of Flight: Stability and Control. John Wiley and Sons, 1996.

[64]   D. Lee, T. Burg, D. Dawson, D. Shu, B. Xian, and E. Tatlicioglu, "Robust tracking control of an underactuated quadrotor aerial-robot based on a parametric uncertain model", in Systems, Man and Cybernetics, 2009. SMC 2009. IEEE International Conference on, 2009, pp. 3187-3192.

[65]   A. Das, K. Subbarao, and F. Lewis. Dynamic inversion with zero-dynamics stabilisation for quadrotor control. Control Theory & Applications, IET, 3(3):303-314, 2009.

[66]   OGATA, KATSUHIKO. Modern Control Engineering. 3rd edition. Prentice-Hall Hispanoamericana, S.A.. 1998.

[67]   M.Murray. Optimization-Based Control. California Institute of Technology.

I want morebooks!

Buy your books fast and straightforward online - at one of world's fastest growing online book stores! Environmentally sound due to Print-on-Demand technologies.

Buy your books online at
www.morebooks.shop

Kaufen Sie Ihre Bücher schnell und unkompliziert online – auf einer der am schnellsten wachsenden Buchhandelsplattformen weltweit! Dank Print-On-Demand umwelt- und ressourcenschonend produziert.

Bücher schneller online kaufen
www.morebooks.shop

info@omniscriptum.com
www.omniscriptum.com

OMNIScriptum

Printed by Books on Demand GmbH, Norderstedt / Germany